Machining
Operation
MACHINER

MW01614702

The most difficult part of a gun to make is the barrel. A knowledge of the conditions under which it will be used, a thorough acquaintance of the principles involved, and sound and accurate machinery are essential before a barrel can be made successfully. Naturally, the sequence of operations and methods used are not identical in different factories, but there are definite stages in its manufacture which all makers must follow. After being centred, the surface of the barrel forging is rough turned to relieve it of outside strains, and briefly, the chief operations following are – drilling, finish turning, grinding outside, fine boring, rifling, lapping, screwing and chambering.

Inspecting

The consideration of primary importance is, that the bore be straight and concentric with the outside, and intermediate viewing and straightening must be performed to ensure this. The tests for straightness in general use will be described later. It is advisable to straighten the barrel before and after drilling and then after turning, the first being obviously a straightening of the outside, and the two latter of the inside, which should be again straightened after reamering. Other straightening operations necessary will depend on the nature of the operations and the strains they impose upon the barrel before they can be designated as an "inside" or "outside" operation. The barrel is officially inspected for the first time in the "fine bored" stage. It is afterwards viewed in the "rifling" stage, and later the accuracy of the screwed end which enters the bolt body is tested; it is again viewed in the first "chambered" stage. The next inspection is to test the fitting of the sight-bed and foresight block, after which the barrel is proved. The barrel is then submitted for view in the "browned and finish chambered" stage. The barrel and body are assembled and viewed in the "breeched up" stage, and finally

Machinery's INDUSTRIAL SECRETS
Selected articles from early issues of Machinery Magazine revealing early manufacturing methods.
Making Rifle Barrels
reprinted by Lindsay Publications Inc – all rights reserved
2002 – ISBN 1-55918-280-6 9 8 7 6 5 4 3

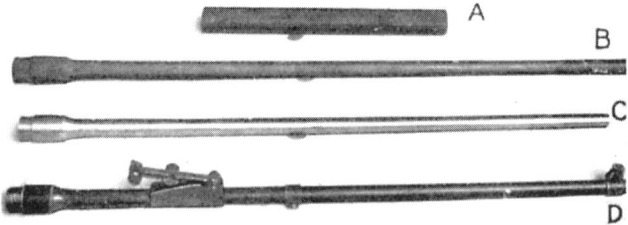

Fig. 1. The Barrel in its Various Stages.

the barrel is inspected after the sights have been fitted. The viewing will be dealt with more fully in the section under the heading of "Gauging and Viewing."

Centring and Turning

A photograph of the barrel in its various stages is shown in Fig. 1, A being the "mould," B the forged barrel, C is a barrel that has been drilled and rough turned, and D is a finished barrel assembled complete with sights and inner band. Nearly 90 operations are necessary to bring the barrel to the latter condition, exclusive of machining the sights. The rough ends of the forging are first cut off by a circular saw, which is mounted on the arbor of a plain geared milling machine with open side. The ends are then centred in a horizontal two-spindle drill. A machine suitable for this is shown in Fig. 2, and is manufactured by Fredk. Pollard, Ltd., of Leicester. The extreme end brackets bearing the spindle are bolted to the bed, and the sliding heads are adjustable along it to any position between the brackets. The spindle and rack feed are similar to the type met with in vertical-spindle sensitive drills. The barrel is held at the breech end in a 3-jaw self-centring chuck which is fixed to the bed of the machine. The muzzle end is supported in a cam–operated 3-point steadyrest. The forging is next

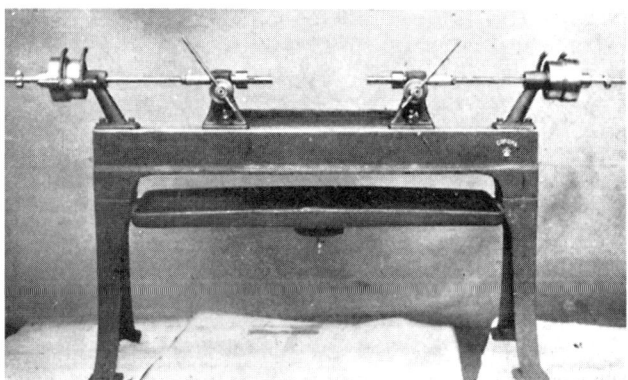

Fig. 2. Spindle Centring Machine: F. Pollard, Ltd.

Fig. 3. Barrel Turning Lathe: J. Archdale & Co. Ltd.

rough turned from breech end to muzzle on the lathe illustrated in Fig. 3. The work which is mounted between the centres is driven by an ordinary carrier and dogs. The power is transmitted to the spindle through the medium of a clutch, which can be seen at A, and is thrown out by the forked lever B; this lever is actuated at the end of the traverse by the end of the saddle coming into contact with the plunger C. The power traverse of the carriage is obtained through a pair of skew gears at D, a worm E mounted on a short shaft, transmitting the motion through a worm-wheel keyed to the end of a long shaft, which can be seen at F, and a half nut on the underside of the saddle which can be disengaged by a lever not shown. At the end of the cut the carriage is wound back by the hand-wheel shown on the front of the carriage, a spur-wheel and rack being the medium employed. The cutter is carried on a bracket which rocks about a centre pivot, the movement of the bracket being effected by the rocker H, one end of which is adjustably connected to the bracket, and the other end is moved up or down by means of a plunger passing through the saddle and riding on the former K screwed to the bed. This former is shaped to conform to the taper of the barrel. A slightly different form of rest is shown in Fig. 4, in which A is the former, B a sliding pad on same, and connected to the arm C, which rocks around pivot D. The cutter is adjusted by the set screw E. A follow rest F is pressed against the barrel by the rocker G and spring H, and is adjusted by a square head set pin and lock-nut K.

In the Pratt & Whitney lathe shown in Fig. 5 a three-roller type backrest is employed, as at A. This rest, which is carried by the bracket B, extending along the rear of the bed, remains in a central position with relation to the barrel, and the tool is guided

3

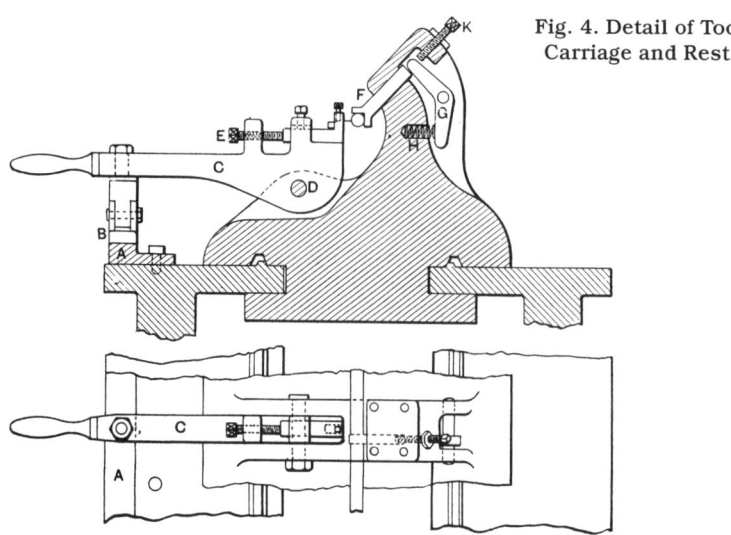

Fig. 4. Detail of Tool Carriage and Rest.

Machinery

by an ordinary taper-turning attachment. The cross-slide carries a bar and roller which bears against the former at the rear of the machine.

Drilling the Barrel

The next operation of importance, drilling the barrel, introduces one of the greatest problems in rifle manufacture, as drilling a hole less than 5/16 inch diameter through 25 inches of tough steel is no mean task. It is not within the scope of this article to discuss at length deep hole drilling, but a brief resume of the principles will be useful. If a long slender rod is put in a drill-chuck and revolved at a fairly high speed the end of the rod will describe a circle round the axis of the spindle. Exactly the same thing occurs with the long drill necessary for drilling deep holes, therefore it would appear that it would be better for the drill to be stationary, and the work, which is much stiffer, to be revolved. A deep hole, or a

Fig. 5. Barrel Turning Lathe: Pratt & Whitney.

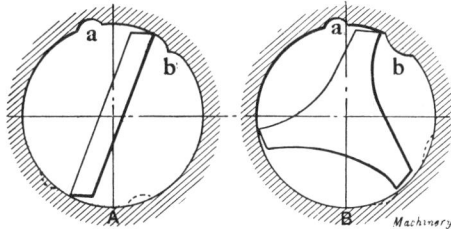

Fig. 6. Drills with 2 and 3 Flutes.

hole of any description where accuracy is required, is best finished by a single-edged tool, which is the usual method used in a lathe for jigs, etc. If a hard or soft spot is encountered when boring with a single-edged tool only the one place is affected. If a tool with two cutting edges is employed, the irregularity caused by the faulty spot is transmitted to the other side of the hole by the spring of the tool. An illustration of this is shown at A in Fig. 6, the dotted parts showing the duplicating of the hard part B and soft spot A. A tool with three edges, as at B, being better supported, would be preferable to one with two, but is more suitable for a reamer. As mentioned before, it is better for the work to be rotated and the drill to be fixed; this is the custom adopted in the drilling of rifle barrels, the outer ends of the barrel being supported in a rest, so that the work runs true. The drill is fixed, inasmuch that it cannot rotate, but is fed forward to the work. If the work is fed on to a rotating drill, the latter when deviating from its true course will take the work with it, thus increasing the deviation. This is brought about by the tendency of the sloping space a Fig. 7 at A, to push the drill still farther out of alignment when the work is fed in the direction of the ar-

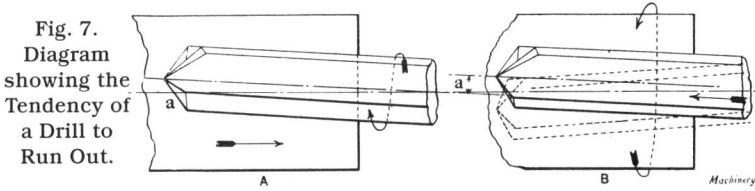

Fig. 7. Diagram showing the Tendency of a Drill to Run Out.

row. When the work is rotated, and the drill is fed forward as at B, the work will carry the drill round in a circle with the radius a when any deviation takes place. The bending action exerted on the drill will tend to force the point back to the axis of the work where no bending occurs. The most satisfactory drill that has been developed yet is a modification of the familiar "D" bit. Fig. 8 shows such a drill entering the work A. B is the cutting edge. Oil passes through the hole C and returns along the groove D carrying the chips with it. It will be noticed that the bot-

5

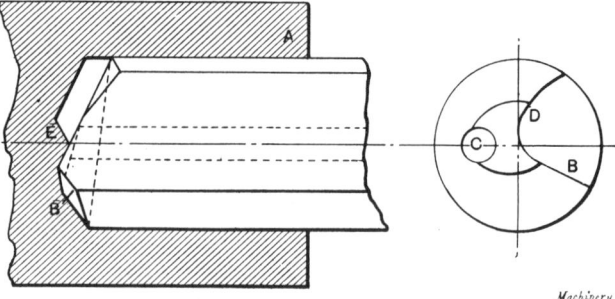

Fig. 8. Barrel and Drill Application.

tom of the groove is taken to the centre line, thus abolishing the disadvantage of the central web as found in ordinary two-lipped twist drills, and making the drill very free cutting. The point of the drill is not on the centre, but a little to one side; this results in a teat E being formed, which supports the drill and greatly assists in producing a true hole. Fig. 9 shows the usual construc-

form a return passage for the oil and chips.

A machine commonly employed for drilling rifle barrels is made by Messrs. Pratt & Whitney, a partial view of which is given in Fig. 10. A chuck is attached to the muzzle end, and the work is rotated at about 2000 revolutions per minute by the catch-plate A, Fig. 11. The end of the drill tube is secured in a

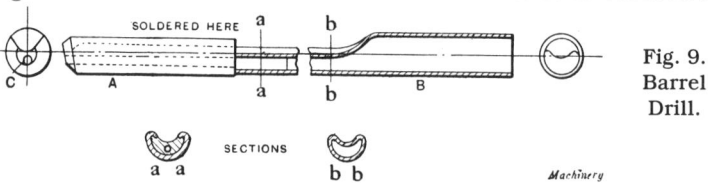

Fig. 9. Barrel Drill.

tion of a barrel drill. A is the drill proper with the oil hole C through it; B is a tube which, as the section shows, is indented for a greater part of its length to

bush B by a set pin, the bush itself being screwed into the nose of the spindle C, which is tightened against a special ball bearing D by the nut at the end of

Fig. 10. Pratt & Whitney Gun Barrel Drilling Machine.

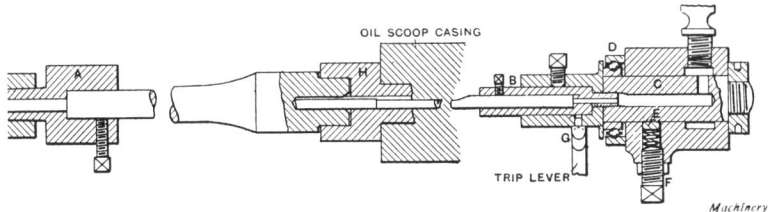

Fig. 11. Barrel Mounted on a Pratt & Whitney
Rifle Barrel Drilling Machine.

the spindle, with just sufficient slackness to ensure the drill being stationary for normal cutting. A brass pad *E* is forced against the spindle by a spring, the ten-

rel drilling machine made by Messrs. J. Archdale & Co., Ltd., Birmingham, is illustrated at Fig. 12. The feed of the drill per revolution of the barrel for drilling a

Fig. 12. Barrel Drilling Machine: J. Archdale.

sion of which is adjusted by a set screw and locknut *F*. If the drill should seize, it rotates, and the small projection *G* in the side of the spindle rotating with it comes into contact with the end of a trip-lever, as shown. This lever operates the belt striking gear and stops the machine. Oil at a pressure of 200 to 400 lbs. per square inch enters at *H*, and passes up the drill tube to the cutting edge, and returns along the groove forcing the chips with it. The oil and chips are deflected at the breech end of the barrel by a scoop *C* in Fig. 10, which also houses the guide bushes *H* (Fig. 11). A very satisfactory bar-

0.303-inch barrel is about 0.003 inch per revolution of the work, which revolves at about 2000 revolutions per minute; the hole is drilled 0.299 inches diameter. The makers of barrel drilling machines supply their machines with individual pumps for the oil service, usually of the gear type.

Some machines are fitted with two pumps, one for each spindle. Pratt & Whitney and Ludwig Loewe machines are so fitted. A very complete outfit is illustrated in Fig. 13. It is a Greenwood & Barley 2-spindle barrel drilling machine fitted with rotary gear pump, oil supply tank, and all the necessary

Fig. 13. 2-Spindle Barrel Drilling Machine: Greenwood & Batley, Ltd.

piping. If a sufficient number of machines is used to warrant it and the construction of the building will allow, one main pump with a common system is preferable to individual pumps, the latter not being powerful enough if a pressure of 500 lbs. per square inch is employed. The lay-out of a very satisfactory system is given in Fig. 14. A horizontal pump of the duplex double-acting type, or a gear-driven 3-throw pump is used. The oil supply is drawn from a large main tank sunk below the pump level. A tank of about 2400 gallons capacity for 60 drills should be used, giving 40 gallons per drill. The oil is pumped through a main pipe A laid in a brick-lined trough running between every few rows, in this case 6 rows of machines. Feeders branch off this pipe to 3 machines on either side of it; each individual machine's oil can be shut off. The feed pipes are also laid in a smaller trough lined with and covered by wood.

The used oil is drained through a pipe running from a sump in a large pan under each machine. This pipe empties into a cast-iron gutter laid under every 3 machines, with sufficient fall on it to ensure the oil running into a large main gutter between the machines. This gutter is laid in a similar trough to the one for the delivery pipe, and also has a fall on it to return the oil through a filter into the supply tank. The swarf is collected from the pans under the machines at intervals and is removed to a centrifugal oil separator, the oil from which is filtered before returning to the supply tank. The arrangement of the piping, etc., given is for a shop having a solid floor; but it would be far more convenient if a shop was used having a basement; the troughs for the various pipes would then be unnecessary, as all the down pipes could pass straight through the floor into mains in the basement. The system outlined could only be carried out in a building with a brick or wood floor. In one having a concrete floor an overhead system for the oil supply could be arranged,

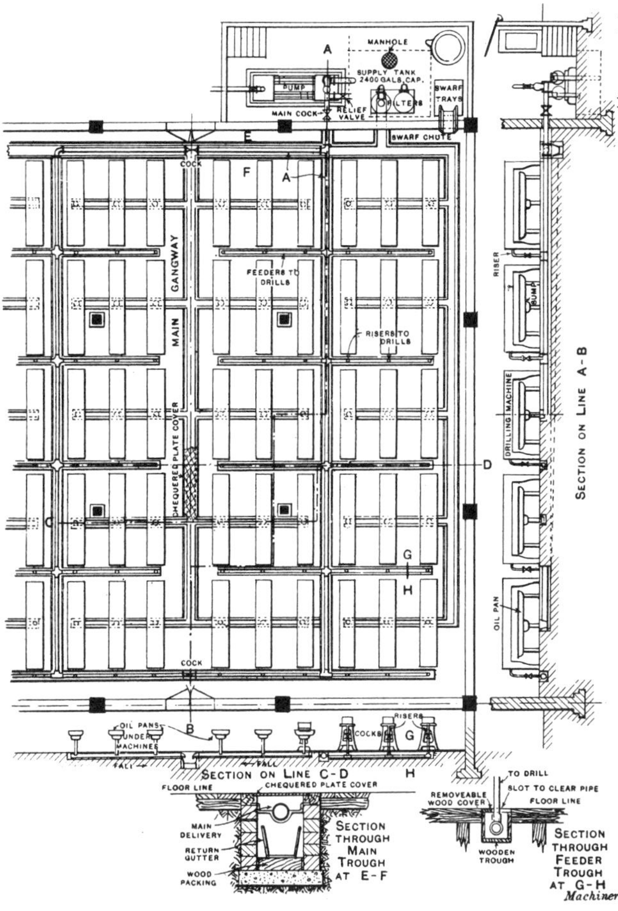

Fig. 14. Lay-Out of Oil System for 60 Barrel Drilling Machines.

Fig. 15. Double-headed Facing Machine: J. Holroyd & Co., Ltd.

9

Fig. 16. Facing Cutter for Barrel.

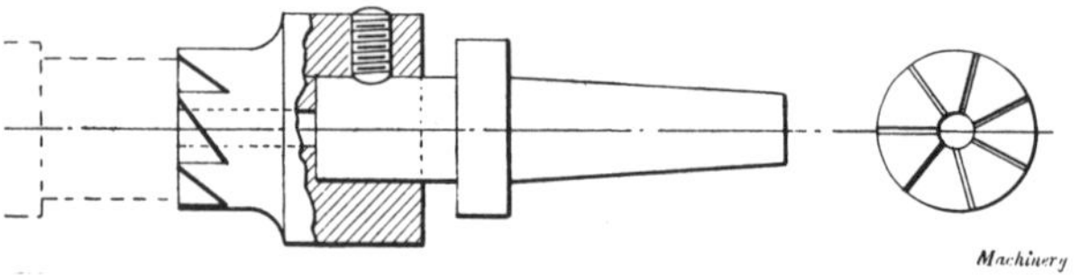

and for the return system guttering or pipes could pass under each line of machines.

The ends of the barrel are faced to length in a double headed facing machine, illustrated in Fig. 15. Between the heads a saddle is bolted to the bed. This saddle carries a slide A, upon which the barrel is mounted, and is moved towards each head in turn by the handwheel shown, operating through a spur gear and rack on the under side of the slide. It is limited in its endwise movement by the stops B. An ordinary facing cutter of the type shown in Fig. 16 is used. The Knox form and radius at the breech end are next machined. A plain lathe is used, of which a plan view is given in Fig. 17. A is the headstock equipped with fast and loose pulleys. B is a special rest to support the centre of the barrel. It consists of a hard wood block C bolted to the bed of the machine with an iron screw to it, the strap having two slots in it to accommodate a wedge E, which secures the removable wood block F. Both blocks are faced with leather where they come into contact with the barrel. Two cutters C and D of the circular formed type are bolted to holders which are held in place by the tool posts E, adjustable along

a tee slot in the cross-slide F. Cutter C is fed into the work first by the handle G operating the cross-slide feed screw, reversing the direction feeds in cutter D. The slide is limited in its movement by the stop H.

Grinding the Barrel

The grinding of the outside of the barrel is the next operation to be performed. The object of the grinding is to give a smooth external finish to the barrel, and what is even of more importance to bring the outside concentric with the bore and true to size without disturbing the straightness of the latter. The plain grinding machine manufactured by the Churchill Machine Tool Co., Ltd., of Manchester, is well adapted for this work. Their 6-inch centre machine taking 36 inches between centres is the one used for grinding the taper of the barrel. The work speed recommended is about 400 revolutions per minute with a table speed of 16 ft. to 18 ft. per minute. The best wheel speed is from 6000 to 7000 ft. per minute. A good wheel finishes about 15 barrels per hour when the amount removed from each barrel is about 0.015 in. to 0.020 in. One truing of the wheel per day is found sufficient. The taper of the barrel is obtained by mov-

10

ing the swivel table of the machine over the required amount. A single steady rest is employed, fitted with a hardened steel shoe to fit the taper of the barrel. A double roller type shoe could be employed if the amount to be removed is small. (Fig. 18.)

For grinding the Knox form a 24 in. by 4 in. ma-

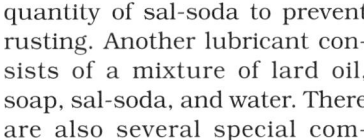

quantity of sal-soda to prevent rusting. Another lubricant consists of a mixture of lard oil, soap, sal-soda, and water. There are also several special compounds obtainable; one of the most satisfactory is "Aquadag," which is used mixed with water. A small quantity of borax, about 1/2 lb. to 10 gallons of wa-

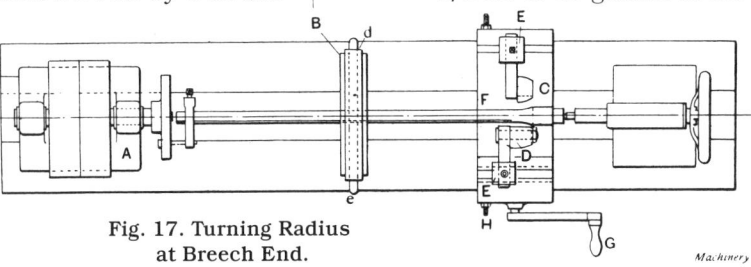

Fig. 17. Turning Radius
at Breech End.

chine is used, with a wheel of 1/2 in. face, 14 in. diameter, the work being traversed across the wheel, which runs at 1500 revolutions per minute, giving 5500 ft. surface speed. The work is revolved at 350 revolutions per minute. Both the grinding operations mentioned should be performed with a plentiful supply of lubricant to prevent heating of the work which would distort the barrel. The most common lubricant is plain water with sufficient

ter, is added to prevent rusting. After grinding, the barrel is straightened, the curve at the breech end first of all being blended into the ground taper with emery cloth. Barrel setting or straightening was formerly the work of highly skilled and highly paid specialists, but improved mechanical methods have considerably reduced the degree of skill required. The old method is still largely in use, as the mechanical means of bending the

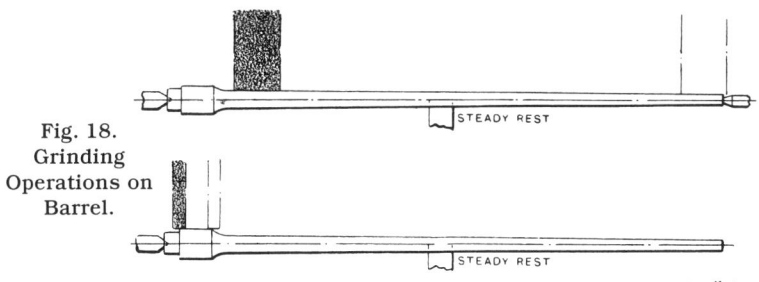

Fig. 18.
Grinding
Operations on
Barrel.

Fig. 19. Barrel Setting Machine.

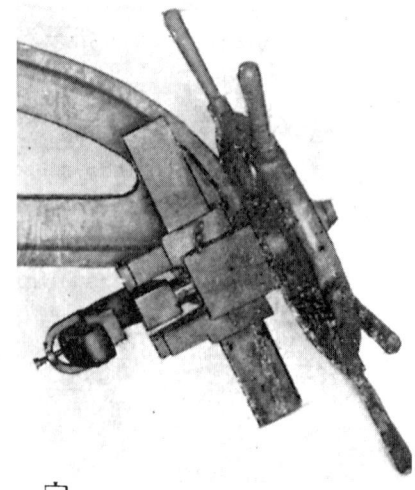

barrel to overcome the inaccuracies has its limitations. The principle of the various mechanical devices for straightening are the same. The barrel is supported on two points on one side, while a single point is applied under pressure equidistant between these points on the opposite side of the barrel, the pressure being usually applied by means of a

Fig. 20. Mechanical Straightening of Barrel.

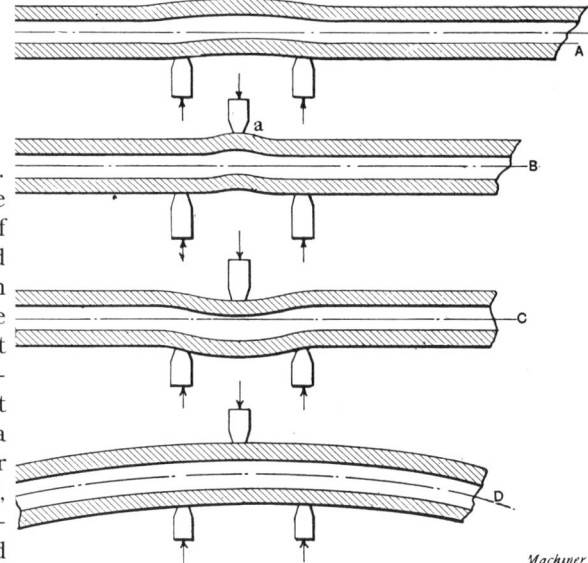

screw, Fig. 19. Some of the limitations of this method are shown in Fig. 20. The bend shown at A will be successfully dealt with, but a short bend or "kink," as at B, when the pressure is applied at A, will assume the shape given at C. The bend shown at D also presents a difficulty, but is more easily overcome than the previous one. Great care and judgment are necessary in using this

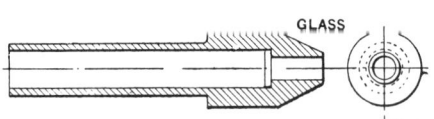

Fig. 21. Centre for Ring Test.

Fig. 22. Reflected Ring
Test.

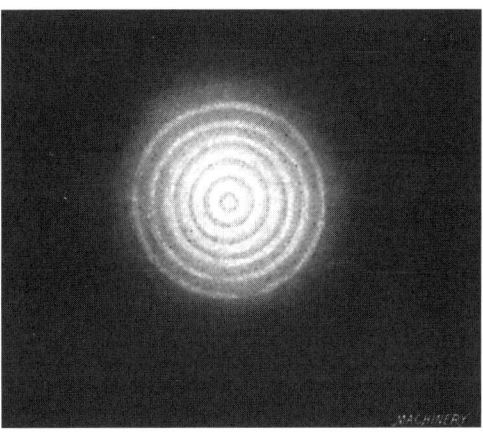

method. When a bend is met with that cannot be treated by the mechanical method, recourse must be taken to the older hand method. Having located the error the barrel is placed across an anvil and is struck in the required direction with a copper hammer.

For testing the straightness of the bore, use is made of the light and shade effects on the smooth interior. One method is to sight through the barrel on to a line placed on the glass of a window. Any curvature in the reflection of this line in the bore will indicate a corresponding lack of accuracy. An extremely rel, but will be eccentric where any inaccuracy occurs. These reflected rings are obtained by placing the barrel between two hollow centres, one of which has a piece of clear glass secured in it. This glass is blacked over with the exception of a thin ring, which is left clear and allows the light to pass through to the bore.

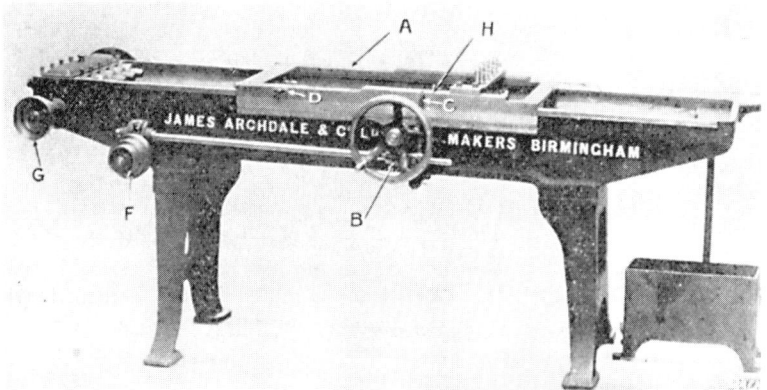

Fig. 23. Barrel Reamering Machine: J. Archdale & Co., Ltd.

delicate test in use depends on the reflections of the muzzle down the barrel, which appear in a series of rings, which should be concentric in a straight bar- A section of such a centre is shown in Fig. 21, and a photograph taken up the bore of a barrel with the centre inserted is given in Fig. 22.

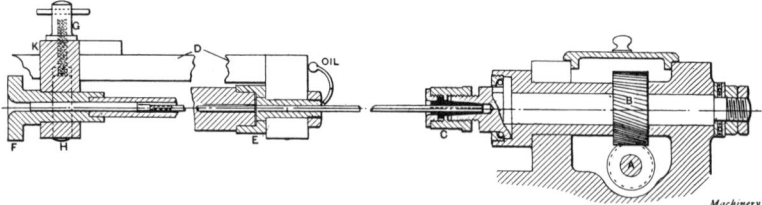
Fig. 24. Barrel Reamering.

Reamering Barrel Hole

The bore of the barrel has now to be brought more closely to the finished dimensions. In the case of the 0.303 barrel in question it is reamered to 0.3025 in. This is performed in a machine of the type shown in Fig 23. Five barrels are reamered at a time. They are mounted between centres fixed in the carriage A, which is moved up or down the bed by the hand-wheel through the medium of a spur-wheel and rack. Feed is obtained through the worm-wheel and worm B, which is disengaged by the trip lever C and stop D. The worm shaft is driven through a pair of skew gears E by the 3-speed cone pulleys F and G. The driving shaft on which the pulley G is mounted carries five skew gears, each gear driving a second gear on the machine spindles. Referring to Fig. 24, A is the driving shaft gear and B the machine spindle gear. A thrust bearing is provided at one end of the spindle and an adjustable ball-bearing at the other. The spindle nose is fitted with a spring collet to grip the end of the reamer, which is tightened by screwing up nut C. D is the carriage of the machine with a set of fixed bushes E and removable bushes F. These bushes are clamped by the nut G, which on being screwed up draws the plunger H, out of which a semi-circular piece is cut against the bush. This also forms a guide bush for the reamer. The centres F are carried in a bracket K, which is adjustable up and down the T slot H (Fig. 23) to accommodate different lengths of barrels. It will be evident from the descriptions and illustrations that the barrels are drawn over the reamers, which are revolved slowly, whilst the barrels do not

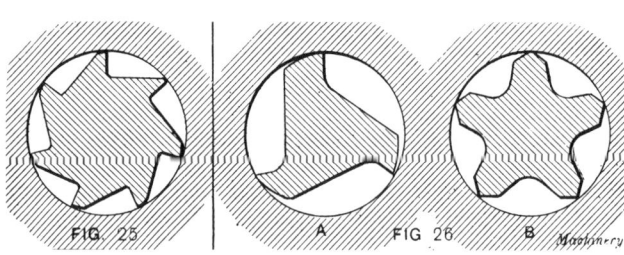

Figs. 25 and 26. Types of Reamers.

turn. A reamer having 7 teeth spaced unequally is used, a helical nick being cut on it to break up the chips; this is illustrated at Fig. 25. It might be remarked here that in all the tools used on the bore of the barrel, and whatever the method employed, provision should be made for the accommodation or removal of the cuttings, as if this is not done it will be impossible to obtain accurate results, the cuttings crowding the tools, which disturbs their alignment and destroys the cutting edges. In America a reamer of the type shown at A in Fig. 26 is used, the amount removed being about 0.003 in. to 0.004 in., the bore being finished by the scraper

shown at B, which removes from 0.001 in. to 0.002 in. The Continental method of finishing the bore is also by a scraping action; the tool shown at A, Fig. 27, is the type used. A very efficient scraper or "bit" is the one given at B. It is ground on the four sides, leaving sharp corners which scrape the bore to the required size. About 0.0005 in. is removed, the bit being packed with a half-round wood spill A; any additional packing necessary is obtained by the paper strips B inserted between the spill and bit. A machine specially built for this job by Messrs. James Archdale & Co., Ltd., is illustrated in Fig. 28. The bits, which are held in a floating chuck, are revolved slowly through the medium of a clutch A on the gear-driven spindles. These clutches are operated by two of the treadles shown: the other treadle B operates the power traverse. The two push-in rods C are for disengaging the power feed to enable the slide to

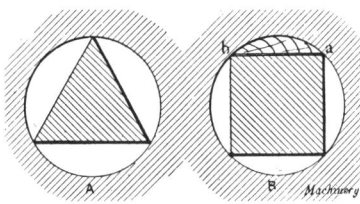

Fig. 27. Scrapers or Bits.

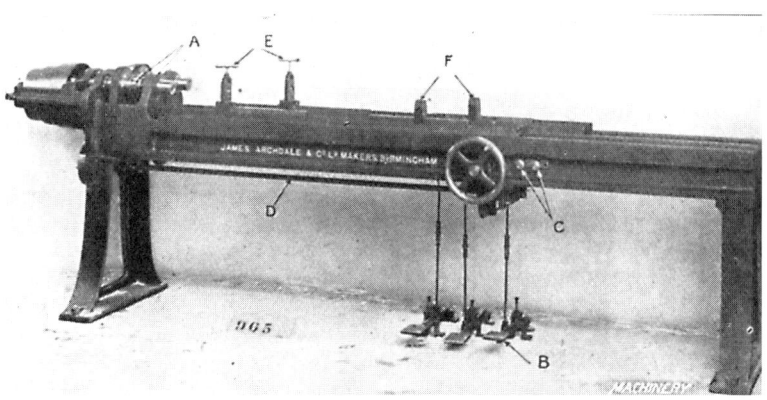

Fig. 28. Machine for Finish-boring the Barrel: J. Archdale & Co., Ltd.

15

be wound back by the hand-wheel. The power traverse is operated through the medium of the shaft D driven by a belt off the main driving shaft, which is equipped with a 3-speed pulley as shown. The breech ends of the barrels are clamped in brackets on the slide by the hand-screws E. The muzzle ends rest in half-round recesses cut in the brackets F.

Rifling and Lapping the Barrel

Rifling, the next operation to be performed, is one of the most delicate to be encountered in the machining of the barrel. The dimensions and limits imposed are given in Fig. 29. It will be noticed that the grooves to be machined are 0.0936 in. wide and 0.005 in. deep. These grooves are cut by means of the special cutter and bar shown in Fig. 30. A is the rod which is drawn through the barrel and mechanically revolved at the same time to give the re-quisite pitch. A cutter box B is pinned to this bar at

C, and carries the cutter D. This cutter is about 1/8 in. thick, and is shaped at E to produce the form of the rifling; the cutting edge is at F. The space at G is to accommodate the cutting chip, which is in the form of a long shaving the length of the barrel, that coils up in this space. The shaving must be removed at the end of each cut. Each groove requires about 6 cuts to be made on it, as the cutter takes approximately 0.001 in. at each cut. When the barrel has made a single revolution, which is necessary to cut the five grooves, the cutter is fed out 0.001 in. by the rod H being screwed outwards, the cutter is then forced up the sloping face K in the cutter box by the spring L. Pratt & Whitney rifling machines are used, the details of which are given in Figs.

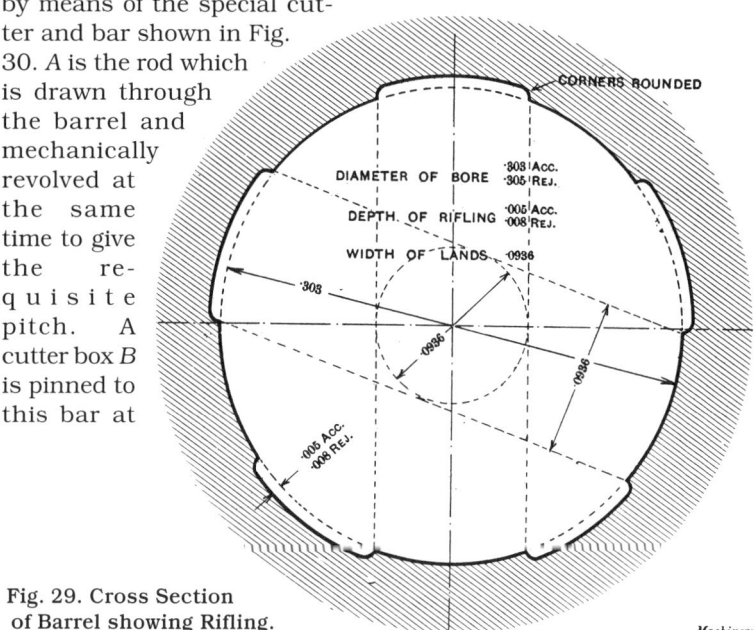

CORNERS ROUNDED

DIAMETER OF BORE ·303 Acc. ·305 REJ.

DEPTH. OF RIFLING ·005 Acc. ·008 REJ.

WIDTH OF LANDS ·0936

·303

·0936

·0936

·005 Acc. ·008 REJ.

Fig. 29. Cross Section of Barrel showing Rifling.

Machinery

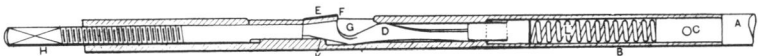

Fig. 30. Detail of Rifling Cutter.

31 and 32. The same reference letters are used in these illustrations. The cutter bar mentioned above is held in a clutch A carried by the sliding head B; this head is alternately moved up and down the bed by a horizontal screw C, the rotations of which are reversed at end of each travel of the head by open and cross belts. The chuck spindle carries a spur gear which engages with a rack secured to the underside of the cross-slide D. On the top of this cross-slide there is a roller which fits in a groove made in the underside of the swivel arm E. As the head B travels along the bed the roller moves along the groove in the arm E which is set at the required angle, and therefore the cross-slide D is constrained to move in a lateral direction, thus rotating a chuck spindle and giving the cutter the required lead. Different leads of grooves are made by setting the arm E to graduations on the quadrant F and then clamping it in position. After the cutter has passed

through the five grooves the square end of the rod H, Fig. 30, enters a square hole in the bracket L the spindle is given a slight turn as the rod enters the hole when the barrel has been turned once. This is effected by a lug on the index disc G coming into contact with suitable mechanism at each revolution of the barrel chuck. The barrel is chucked in the arrangement shown in Fig. 33. The Knox form at the breech end fits in a hardened steel bush K and is clamped by a set screw L. A bush M slides over the muzzle end and fits in the hollow spindle N. Keyed to this spindle there is a gear O meshing with a rack P, which, at the end of each stroke of the cutter, revolves the spindle an amount sufficient to cut the next groove. Also keyed to the spindle is an index disc G with five grooves in it. A plunger Q enters these grooves and locates the spindle for cutting the rifling grooves. This plunger is lifted for indexing by the special cam R.

After rifling, the barrel has

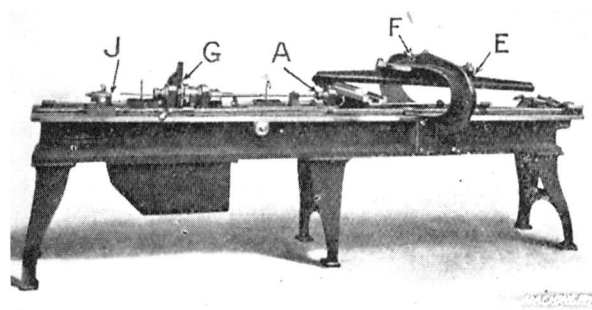

Fig. 31. Pratt & Whitney Rifling Machine.

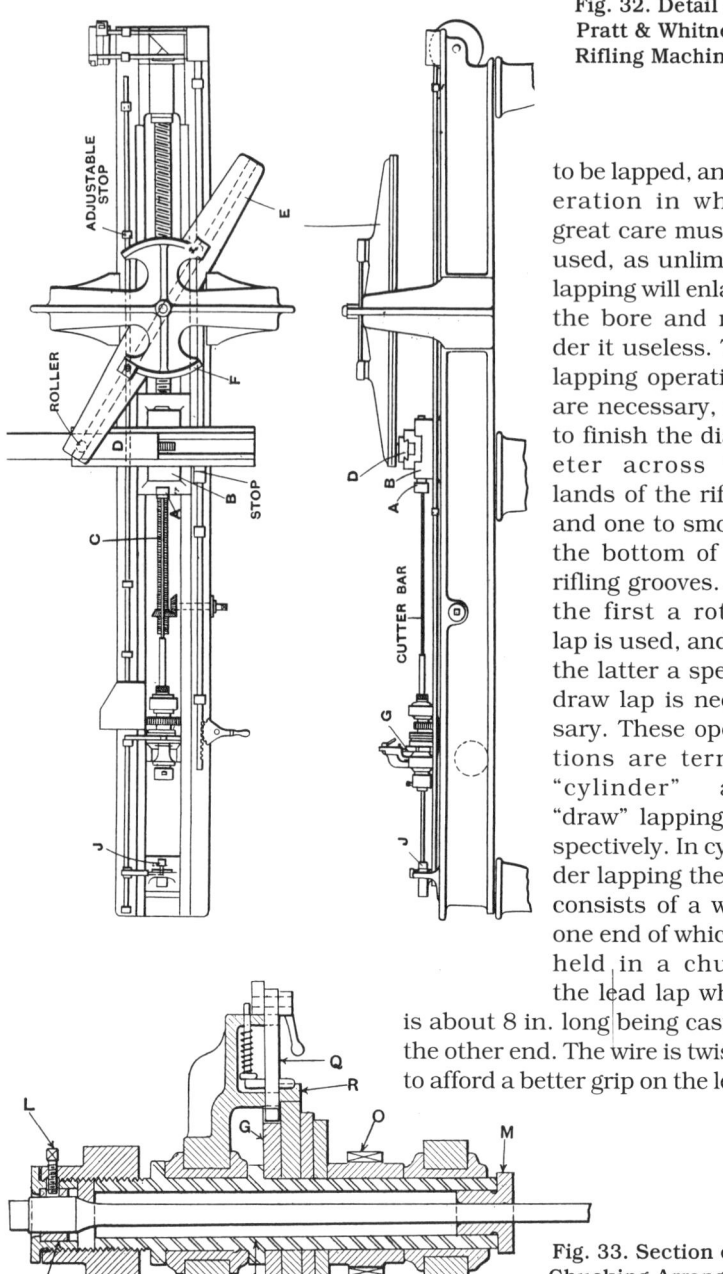

Fig. 32. Detail of
Pratt & Whitney
Rifling Machine.

to be lapped, an op-
eration in which
great care must be
used, as unlimited
lapping will enlarge
the bore and ren-
der it useless. Two
lapping operations
are necessary, one
to finish the diam-
eter across the
lands of the rifling
and one to smooth
the bottom of the
rifling grooves. For
the first a rotary
lap is used, and for
the latter a special
draw lap is neces-
sary. These opera-
tions are termed
"cylinder" and
"draw" lapping re-
spectively. In cylin-
der lapping the lap
consists of a wire,
one end of which is
held in a chuck,
the lead lap which
is about 8 in. long being cast on
the other end. The wire is twisted
to afford a better grip on the lead.

Fig. 33. Section of
Chucking Arrange-
ment, Pratt & Whitney
Rifling Machine.

This lap is given a reciprocating motion up and down the barrel, which is stationary, at the same time the lap is revolved at about 200 to 300 revolutions per minute. Draw lapping is accomplished by means of a somewhat similar lap, the difference being that grooves are cut in the lead the same pitch as the rifling and deep enough to clear the lands. Across section of a barrel with lap in position would appear as Fig. 34. To make the lap the twisted end of the wire is placed in the barrel with tow wrapped round it to form a stopping, molten lead is then poured in the barrel and sets in the grooves of the rifling. The lap is relieved to clear the lands by drawing it through a barrel in which is inserted a cutter the same width as the lands and adjusted a little deeper. An idea of this device may be gathered from Fig. 35. The laps are used with a mixture of emery and oil. The illustration (Fig. 36) shows a machine made by Greenwood & Barley, of Leeds, for draw lapping. A reciprocating motion is given to the slide A by means of a rocker B. A connecting rod C transmits the motion from a similar rocker in the bottom of the machine, which is given the necessary movement by means of a crank motion which is transmitted from the driving-wheel, seen at the side of the machine, by gears. Five barrels are lapped at a time; the muzzle ends resting in coni-

TWISTED SQUARE END OF ROD RELIEVED TO CLEAR LANDS OF RIFLING

LEAD LAP

Machinery

Fig. 34. Section through Barrel showing Lap in Position for Lapping Rifling Grooves.

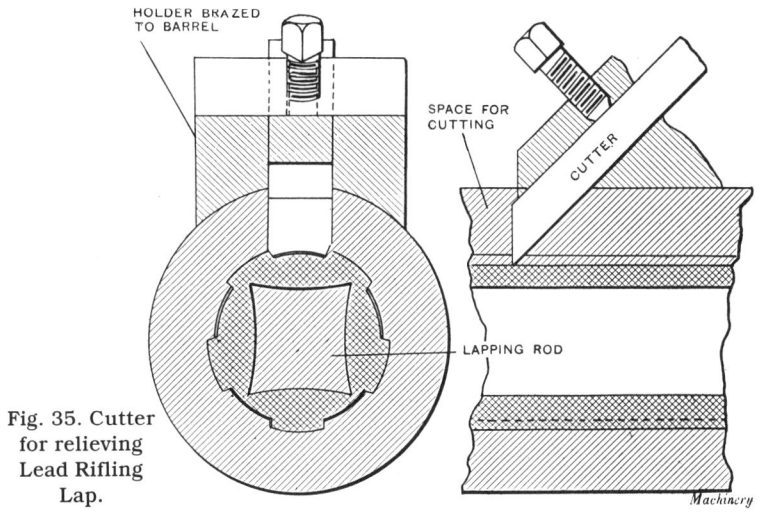

HOLDER BRAZED TO BARREL

SPACE FOR CUTTING

CUTTER

LAPPING ROD

Fig. 35. Cutter for relieving Lead Rifling Lap.

Machinery

Fig 36. Draw Lapping Machine: Greenwood & Batley, Ltd.

cal recesses in a plate *D*. The breech ends are clamped down by the plates *E* and nuts shown, a spring being fitted under each plate. The laps are held in the chucks *F*. Each chuck spindle has a gearwheel on it which meshes with a gear encircling the former *G*, the latter having a groove in it the same pitch as the rifling. For cylinder lapping the fixed former is replaced by a rotating shaft, a sliding key transmitting the rotation through the gear to the chuck spindles. The same principles are used in the horizontal lapping machine shown in Fig. 37, made by James Archdale & Co. In this case the laps are passed through the barrel and are held at each end in chucks *A* and *B* at the ends of the bed. An important difference on this machine is that the barrels are given the reciprocating motion, but a very similar crank motion is used to move the slide, which in its travel rotates the former *C* at the side of the machine and revolves the laps through gears on chucks *A*. The same method of conversion for cylinder lapping is possible as that on the machine previously mentioned.

Screwing and Chambering

The barrel has next to be screwed on the breech end. To prepare it for screwing, the

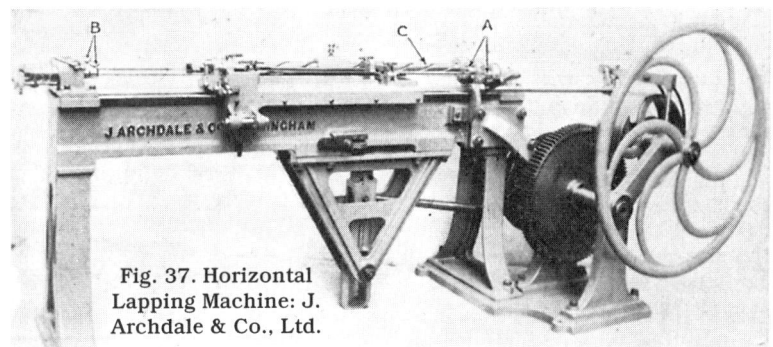

Fig. 37. Horizontal
Lapping Machine: J.
Archdale & Co., Ltd.

breech end is brought to the correct diameter and length in a Holroyd capstan lathe of the type shown in Fig. 38. This machine is built specially for this operation. The turret is mounted on a cated in position for cutting by a removable plug, and is fed towards the spindle by the handwheel shown at the end of the bed. Referring to Fig. 39, the first operation is to face the breech

Fig. 38. Capstan Lathe for Breech End: J. Holroyd & Co., Ltd.

cross-slide, and can be pushed back by hand to the rear of the machine to enable the barrel to be withdrawn. The turret is located end of the barrel to length. It is turned to the correct diameter by the box-tool B, tool D chamfering the corner at the same time.

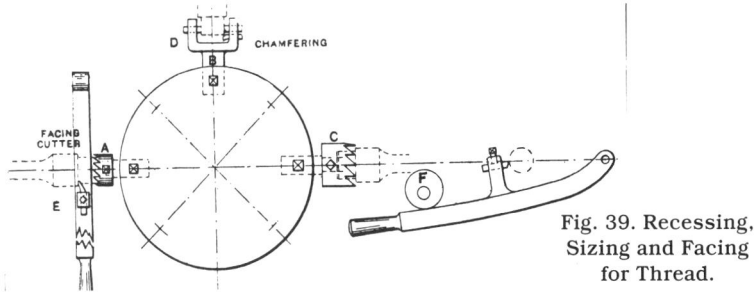

Fig. 39. Recessing,
Sizing and Facing
for Thread.

21

The hollow mill C then finishes the diameter and the face of the shoulder. A recess is formed for the end of the thread to run into by the cutter E, mounted on an arm pivoted at the rear of the bed. This arm is held against a cam F rotated off the headstock, the result being a groove round the barrel corresponding to the shape of the cam. This arrangement will be clearly seen on the machine in Fig. 38. The barrel is held in a chuck very similar to the type used on the machine for thread-milling the breech. The arrangement of this chuck is to the barrel. During this tightening the breech of the barrel butts up against an adjustable stop secured to the front of the headstock. A hardened plug enters the muzzle end, which is guided during insertion by the internal taper at G. The hardened and ground spindle H is carried in adjustable phosphor-bronze bushes, and has the guide-thread cut on it at I, which passes through a phosphor-bronze nut, also adjustable for wear. To the spindle is keyed a hand-wheel K which has four recesses in the base. These re-

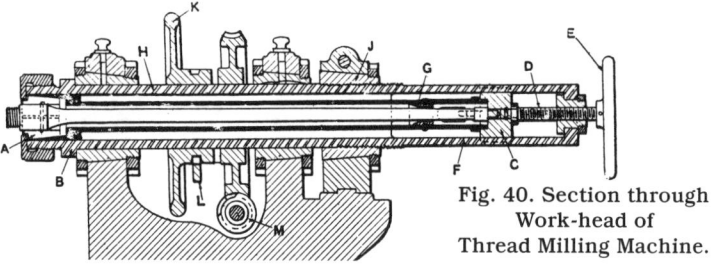

Fig. 40. Section through
Work-head of
Thread Milling Machine.

seen in Fig. 40. The barrel is held by the Knox form in four split jaws A fitting in a coned recess in the bar B. The rear end of this bar is screwed into a sliding piece C actuated by the screw D and hand-wheel E. It will be obvious that when the hand-wheel is turned the coned face at the front end of B, sliding over the split jaws, will tighten them on cesses engage in a lug on a bar L, which is held in the position shown to prevent the spindle turning whilst the chuck is being tightened. A driving worm-wheel for revolving the spindle is also secured to the spindle and

Fig. 41. Thread
Milling Machine:
J. Archdale & Co.,
Ltd.

22

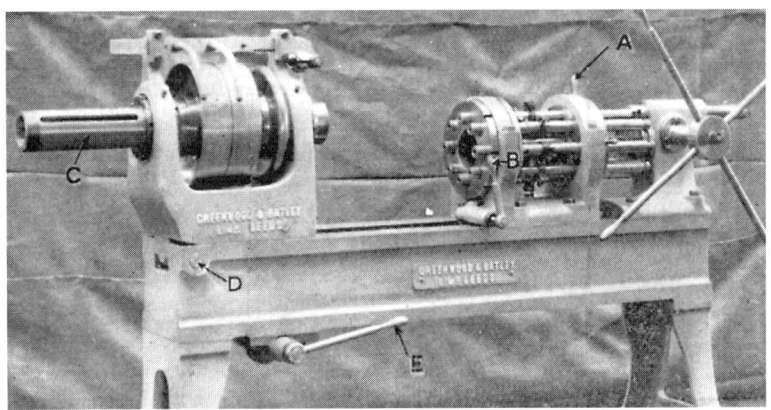

Fig. 42. Chambering Machine: Greenwood & Batley, Ltd.

is rotated through the worm *M*. This worm is mounted in an eccentric sleeve, and is brought into engagement with the worm-wheel by a quarter-turn of the handle *A* (Fig. 41).

After the thread has been sized the bore at the breech end has to be enlarged and machined to take the cartridge. It is very important that the axis of the cartridge chamber be coincident with that of the bore, therefore care must be taken in machining, and an efficient and accurate machine must be used. Machines with several tool-holders are necessary for this work, so those of the capstan type are used. Capstans of the orthodox type – that is, those in which the axis of the capstan is at right angles to the spindle and bed – no matter how carefully they are made, will develop little inaccuracies which would be fatal to the work referred to. Fig. 42 illustrates the kind of machine used. It is made by Greenwood & Batley, Ltd. The axis of the tool

spindles is in alignment with the hollow work spindle and barrel. These spindles, of which there are eight, are fed forward by means of the large capstan shown. The spindles are carried in a revolving cage which is indexed round by the lever *A* and located by a plunger operated by lever *B*. The cage revolves in a head which is bolted securely to the bed of the machine, and is not moved during cutting. The barrel is held in a chuck very similar to the one described for the screwing, and is passed through the hollow spindle *C*. The work-head is pivoted at *D*, and the front end is raised to enable the barrel to be withdrawn by a cam-lever *B*. In the chambering machine made by Pratt & Whitney the muzzle end of the barrel is held in a spring collet chuck in the spindle nose, the spindle head being an integral part of the bed. The Knox form rests in a semicircular steady. The type of machine used on the Continent is illustrated in Fig.

Fig. 43. Another Type of Chambering Machine.

43. The tools are held loosely in a square hole in the bracket A on the tool-slide; a full floating effect is thus obtained. The cutters for roughing out the chamber are shown in Fig. 44. Cutter A is for roughing the taper of the cartridge case, B roughing for the neck of the ease, C reamers for the neck, and D the taper; E reamers the bullet lead. The chamber is finished by hand by the reamers shown in Fig. 45. Reamers A and B finish the taper of the ease and the small radius at the commencement of the neck. C reamers the neck and D and E finish the bullet lead. F is for putting a small radius on the corner of the chamber. All the tools mentioned are used in a hand-brace, and are pro-

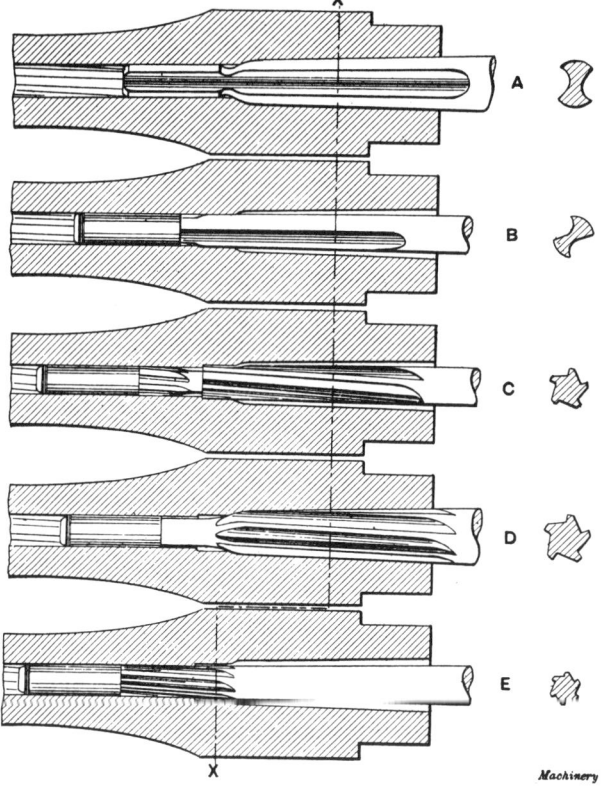

Fig. 44. Operations on the Cartridge Chamber.

Machinery

Fig. 45. Reamering Cartridge Chamber.

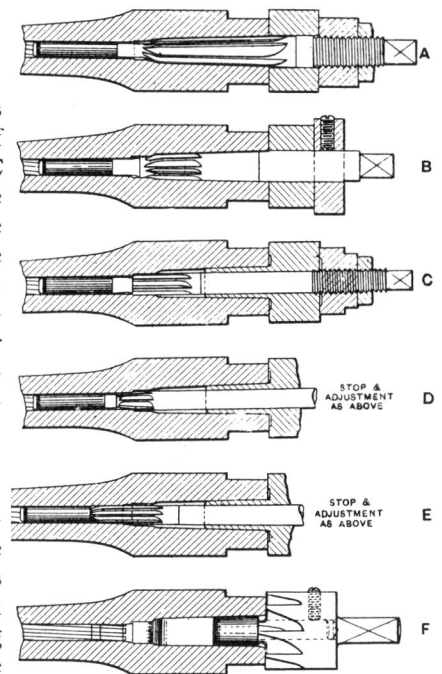

vided with adjustable stops which gauge from the face of the breech. Tools *C, D* and *E* run in bushes which fit the taper of the chamber. The finished dimensions of the chamber are shown at Fig. 46. Previous to the finish-reamering of the chamber several minor machining operations have to be performed.

Minor Operations

The groove for the extractor has to be cut and the flat on the Knox form is milled and then ground in the machine shown in Fig 47. The keying for fixing the front sight block is slot-milled in an Archdale vertical slot miller. The flat is ground with a cup wheel which is mounted in a reciprocating slide, the motion being produced by a crank disc driven by a worm. The back sight is then fitted to the barrel and the hole for the spring screw is drilled, an operation calling for little comment. The front sight block is next fitted, after the inner band is put on. The fixing

pin-holes in the back sight bed and front sight block are drilled in a horizontal 2-spindle drill of the same type as illustrated in Fig. 2. The barrel is mounted in a jig secured to a baseplate bolted to the bed of the machine between the spindles. This machine is also used for drilling the axis pin-hole in the sight bed for the sight leaf. The position of the sight block and bed are gauged

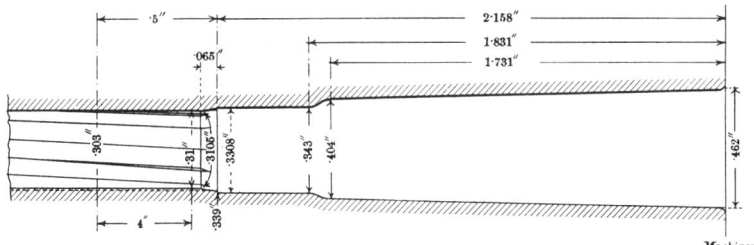

Fig. 46. Details of Cartridge Chamber.

Machinery

Fig. 47. Machine for grinding Flat on the Knox Form: J. Holroyd & Co.

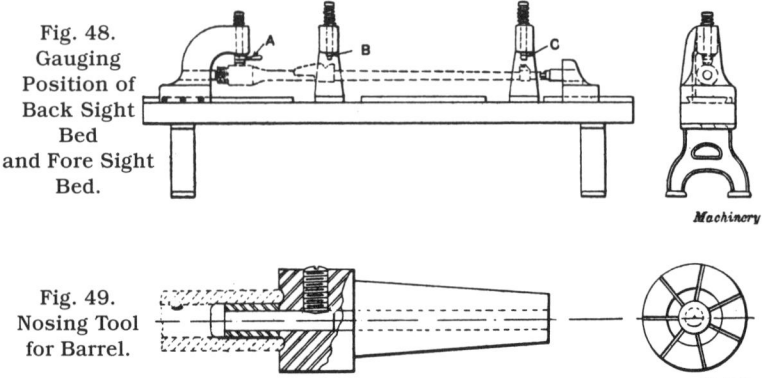

Fig. 48. Gauging Position of Back Sight Bed and Fore Sight Bed.

Machinery

Fig. 49. Nosing Tool for Barrel.

Machinery

by a special jig shown in Fig. 48. A taper plug entering the chamber, and a parallel one the muzzle end, support the barrel, and a cam-operated plunger A locks it in position and locates it from the flat on the Knox form. The spring plungers B and C, a push fit in the brackets, gauge the back sight bed and front sight block respectively. A locating piece on plunger B enters be-

Fig. 50. Surface Gauge for Checking Sight Elevations.

tween the ramps on the sight bed. On plunger *C* the locating piece is cut to fit on the sides of the front sight block.

At this stage the barrel is proved, together with the action. It is then stripped, and the barrel is prepared and then browned. The proof and browning will be further dealt with in separate sections. The barrel is cleaned and the muzzle end is radiused or "nosed"; the cutter shown in Fig. 49 is used in a similar type machine to that used for facing, the difference being that it has one head only, instead of two. The body that has already been fitted to a barrel is now screwed to the same barrel, the fore sight is fitted to the sight block, and the back sight leaf and spring are assembled to the bed on the barrel. The ramps of the sight bed have to be corrected to bring the sight leaf to the exact height necessary for the various ranges. This is performed by filing the ramps at various points along their length that correspond with settings of the sight slide to give elevations. These elevations are checked by means of a special surface gauge shown at *A* in Fig. 50. This gauge has steps on it, each step representing a different elevation of the sight slide. The barrel is supported by a plug entering the cartridge chamber and another the muzzle; the surface gauge is used on the bed of the fixture. The fixture illustrated in Fig. 50 is for viewing or "needling"; the accurately fitting spring plunger *B* has-a flattened needle at its lower end; on the plunger being depressed, the needle should enter the U on the sight wind gauge. The exact figure of the ramps is also tested with a special surface gauge.

Drilling, Reaming, and Straightening Rifle Barrels
By W. H. A,
MACHINERY MAGAZINE - JUNE 1, 1916

Rifle making experts in general concede that the most difficult part of a rifle to make is the barrel. A rifle with a barrel which is untrue is worthless; hence, the greatest care in manufacture is necessary in order to ensure that this portion be as nearly correct as it is possible to make it. Should the interior of a barrel be out of alignment, the rifle would be unreliable and could not be depended upon to give accurate results under all conditions.

It can readily be seen from the foregoing that gun barrel making is a business in itself. As a matter of fact, there are factories where only gun barrels are manufactured and where no other gun parts are made, although there are no factories making other gun parts (except sights) exclusively. All the important operations on the barrel are carried on in the barrel depart-

Fig. 1. Cutting Off Stock for Rifle Barrels.

ment, although some of the minor operations are performed in other parts of the factory. The operations done in the barrel department are: cutting off the stock, outside straightening, centring, drilling, spotting, turning, reaming, straightening, grinding, finish-reaming, rifling, etc. All these operations are important and must be done correctly or else the barrel will never "make." Threading, sight-seat milling, slot-cutting, rifling, polishing, browning, and other operations, as well as proving under powder tests, are done in departments not connected with the barrel department proper.

Rifle Barrel Material

While rifle barrels were formerly made of mild steel exclusively, the modern military rifle barrel is made of a much tougher grade of steel, owing to the more exacting requirements of the modern powders as contrasted with the strength needed when only black powder was used as an explosive. Thus the nickel steel barrel has come to the front and no modern military rifle barrels are made of mild steel at present. The stock from which the barrels are made usually comes to the factory in long bars which vary in length and diameter according to the style of barrel to be manufactured. The general specifications for rifle barrel steel at the present day are: carbon, 0.50; manganese, 1.00; silicon, 0.30; phosphorus under 0.015; sulphur, under 0.015.

Cutting Off Stock

The first operation to which the bars are subjected is cutting off, the bars being placed two or three at a time in a cutting off machine which is supplied with a gang of circular saws. Fig. 1 shows an arrangement for cutting off rifle barrels to their proper length, due allowance being made on the length to provide for errors that may be made in machining. This is a multiple saw of the Higley type, having blades 12-1/2 inches in diameter and 9/32 inch thick. The bars are clamped in place and are fed directly against the teeth of the revolving saws, the feed of the work being 5/8 inch per minute and the speed of the blades 10 r.p.m., which is equivalent to a cutting speed of 32 feet per minute. By means of a pump, a stream of oil is constantly directed on to the saws while the cutting operation is in progress, thus keeping them properly lubricated during the cutting-off operation. It is possible to arrange the saws so that barrels of various lengths can be cut off at the same time. A gang of eleven saws can be easily tended by one man with a helper.

Upsetting the Stock

After the stock has been cut into the proper lengths for barrels, it is upset at the breech end in order to bring it to the correct shape, after which it is ready for annealing. Two methods are used in upsetting the stock, viz., forging in a regular forging machine and a process of rolling.

In the factories where the barrel lengths are forged, the ram of the machine thickens and forms tile butt of the barrel while the barrel is held lengthwise between two split dies. When the barrels are shaped by rolling, a pair of rolls is employed, these rolls having about six sets of various sized grooves, which are deep at the beginning and diminish as they progress round the rolls. As the deep cavity of the first groove in the roll comes round to the operator, he thrusts the end of the barrel stock into it and partially shapes it. By following this procedure with each set of grooves successively, the barrel is finally brought to its proper shape.

Annealing Barrels

After the barrels have been upset they are taken to the annealing furnace, where they are slowly heated to the critical temperature, which is dependent on the grade of steel being treated. They are then removed, packed in lime and allowed to cool slowly.

Outside Straightening

Before the barrels are sent to the drilling machines, they are looked over by a workman and any that are crooked on the outside are straightened by him. This process is called "outside straightening," and while it is important, it does not necessarily require an expert workman, as anyone of ordinary intelligence can learn to do it in a short time. If the straightening were improperly done and the barrels

Fig. 2. Butt Turning Breech End of Barrel on Shaving Machine.

went to the drilling machines in a crooked condition, the result would be that the drills, in passing through the stock, would come out at one side instead of in the centre, or that the walls of the barrel would be thick in some places and thin in others. Either of these conditions might result in disaster when the barrel was turned down to its proper size and shape in the turning lathe. In other words, a barrel which is not properly straightened on the outside might not "make" in the subsequent operations.

Butt Turning

After annealing, the barrels are taken to the butt-turning machine as indicated in Fig. 2. In this operation, the barrel is placed through the spindle of a shaving machine and the butt end held in a split chuck while the end is turned down and made ready for the drilling operations. Some factories also centre the butt end in order that the drill may enter the stock directly in the centre when the drilling operation is started, while in other factories no centring is done.

Drilling Rifle Barrels

The drilling of a rifle barrel is an especially important operation for the reason that if a barrel is not properly drilled, it almost invariably finds its way to the scrap heap. In order, then, that this may not happen it is absolutely necessary that the drills used be perfect, both as to gauge and cutting capacity. The man who keeps the drills in proper condition, although he may not operate the machine, is called the "driller", and he must be a man having more than ordinary skill and intelligence. Highly skilled drillers who keep the drills in first-class condition are difficult to find.

Two types of drilling machines – horizontal and vertical – are in use today. One of the larger companies uses a vertical drilling press for barrel drilling, which was, perhaps, until recently, the most-perfect press of its kind, and which is manufactured at the factory where it is used. These drilling machines have several secret attachments connected with them not found on other styles of drill presses.

Fig. 3. Rolling in Drill Shank
Groove on Planer with Special
Fixture.

immense oil system is located, consisting of a series of reservoirs holding many hundreds of gallons of oil which is forced by pumps of high pressure through the whole drilling system. The pumps connected with this oil system develop pressure of 800 lbs. or more, by which the oil is kept in constant circulation, passing through each drill and finally back to the great central reservoir in the basement. The drill chips which come from the centre of the barrel are automatically carried out by the flow of oil and deposited in receptacles which are connected with the drilling system.

Gun barrel drilling is a difficult operation on account of the

One of these attachments is a motor contrivance by which the drills can be speeded up from 1300 revolutions per minute to 3000 without necessitating a change of belting. When drilling nickel steel barrels, or 0.45 carbon steel, the speed at which the drills are run is about 1760 revolutions. For soft steel barrels, however, the highest speed obtained is about 3000 revolutions. The time necessary for drilling a high-grade steel barrel is from forty-five to sixty-five minutes, while the maximum speed at which soft steel barrels are turned out is about twenty minutes, according to the length of the barrel.

Under the drill presses at this factory, in the basement, an

Fig. 4. Gauging Accuracy of Gun
Barrel after Reaming.

31

Fig. 5. Pratt & Whitney Gun Barrel Drilling Machine.

extreme length of the hole as compared with the diameter of the drill and also because of the toughness of the steel in which the drilling must be done. In either horizontal or vertical drilling machines, the barrels are revolved and the drills held stationary. The vertical machines are a more recent development and are considered to have several advantages over the horizontal types, such as reduced floor space and better removal of chips. The horizontal type of machines is best known and will-therefore be described here. Fig. 5 shows a Pratt & Whitney gun barrel drilling machine, while Fig. 6 shows a closer view of the work on the machine.

On this type of machine, two barrels are drilled simultaneously, there being two spindles, each of which is driven independently from an overhead belt. Beneath the spindles are worm-gears that carry motion to the lead-screws outside the two walls of the machine. These lead-screws can be seen in Fig, 6, and from each of these screws the corresponding carriage is operated. The feed of the carriages may be engaged or disengaged by hand-levers, and the disengagement is also automatically effected by a set of stops, as shown in Fig. 5. The rate of speed on a gun barrel drill depends on the quality of the steel which is being drilled. The barrels are usually drilled at a speed of about 1760 revolutions, as previously mentioned, and the rate of feed is about one inch per minute, which is about the maximum rate. The method of driving the barrel is by a contact on the muzzle end against the cen-

Fig. 6.
Enlarged
View of
Drilling
Machine,
showing
Carriages.

tre of the spindle, while the other end is supported in a bushing as indicated in Fig. 6. The pump pressure, by means of which the oil is circulated through the drill when drilling, is of great importance, as this pressure must be sufficient to work out the chips along the passage provided for them so that they will not clog or cause other trouble. The shank of the drill rod is securely fastened in the chuck on the drill carriage at the end of the machine, while the tip or cutting end of the drill passes through a bushing at the centre of the machine.

Shape of Drills

The shape of the drills used by all the large manufacturing concerns is practically the same. The shank to which the drill tip is brazed is the same shape as the drill, that is, half round, with a deep groove running from one end to the other, and having a crescent shaped hole which passes completely through the shank and drill tip. The method

by which the groove in the shank is obtained is by placing the shank, which is a round thin-walled tube, in a special fixture and rolling in the groove under the pressure of a revolving wheel as shown in the illustration; Fig. 3. This roll is forced down on the tube as the planer bed travels, thus compressing the metal and forcing it to take the form shown. The method of obtaining the crescent shaped hole in the drill tip is not the same as that employed on the shank. This tip is from 3-1/2 to 4 inches in length and is made of the best grade of high-speed steel. It is drilled through the centre in its rough state before it is shaped up, the hole extending from end to end. The tip is then placed in a die and drop-forged to produce the groove from end to end in the tip. The displacement of the metal caused by forging in the groove causes the round hole to become crescent shaped. The tip is then shaped up and brazed to the end of the shank, after which the cutting end is ground to its

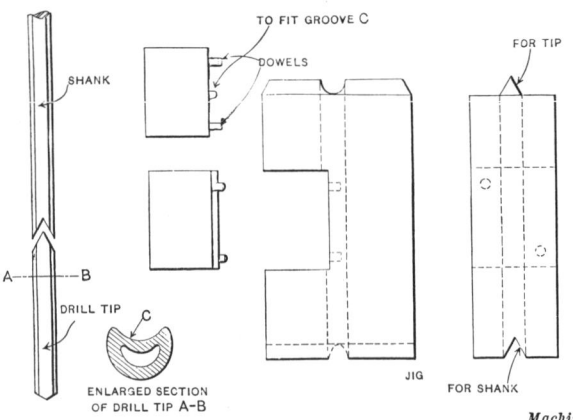

SHANK

TO FIT GROOVE C

DOWELS

FOR TIP

DRILL TIP C

A —— B

ENLARGED SECTION
OF DRILL TIP A-B

JIG

FOR SHANK

Fig. 7. Jig for Filling Drill Tip and Shank.

proper shape by the operator. The method of forming the end of the drill tip and its tube is shown in Fig. 7, which represents a filing jig, in which the shank and tip are filed to correspond to the angular surfaces on the jig.

In grinding the drill tip, the practice varies somewhat in different factories, some claiming that the drill should be ground slightly below centre so as to drill (as it is called) "for a wire," while others believe in grinding the drill tip on centre in the usual man-

ner. When, drilling "for a wire," a slender thread of steel is left at the centre of the barrel, and when the end of the drilling is reached a disc like that shown at A in Fig. 8 drops out. Care should be taken in the amount of feed used on the drill so that the chips may come out in such a way that they will be readily carried out through the groove in the drill shank without any tendency to clog. It is much better to have the chips like those shown at B in the illustration,

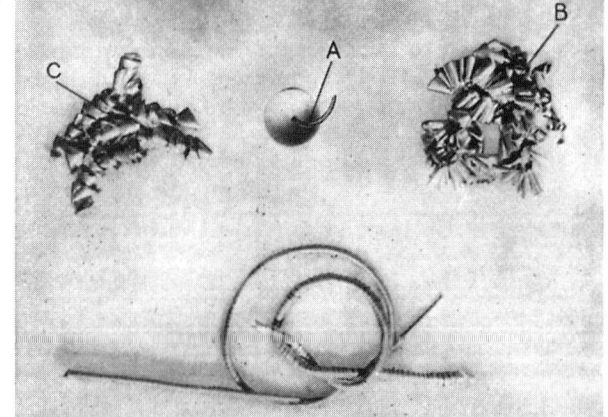

Fig. 8.
Samples of
Turning and
Drilling Chips
and a Disc A
removed when
"drilling for a
wire."

which are fine and can be reduced to powder between the fingers, than to have them long as shown at *C*, for the reason mentioned. Medium grade oil is generally used in drilling barrels, although some manufacturers use a high grade of oil. It has been found, however, that the medium grade answers the purpose just as well as the higher quality, and it is stated by some manufacturers that even better results are obtained by its use.

To obtain the best results in drilling rifle barrels, either of the military or sporting type, the very best judgment is necessary. When the drills are over-speeded or crowded, the result is a barrel defaced with drill rings and gouges from one end to the other. In addition, the operations which follow the drilling are retarded by attempts to save the barrels from the scrap heap. Judgment must therefore be used so that the drilling machines are manipulated in such a way as to produce the greatest amount of good work possible without over-speeding.

Varying Qualities of Steel

The quality of steel used in rifle barrels has a far-reaching effect on the drilling processes, and brands of steel which may have been perfectly satisfactory at one time are often the cause of trouble at others. Single lots delivered at factories sometimes contain such variable qualities of steel that the greatest difficulty is experienced in working the stock up into barrels. Thus,

in some shipments, part of the stock will run good while other portions will create trouble from the cutting-off process to the finished article. Soft and hard spots will be found in the stock and the drilling is therefore much handicapped, while the other processes such as straightening, reaming, grinding, etc., are also interfered with. In the working up of such stock, an inferior article is likely to be produced on account of the variable conditions under which the work is manufactured.

Reaming the Barrels

The first reaming operation usually comes just after drilling, but this is not always so, as some factories turn the barrel both before and after reaming. The tendency to-day is to do away with as many operations as possible and at the same time obtain the best results. Instead of using a medium or low grade of oil in the reaming operations, the highest grade of lard oil is used. The steel used for the reamers varies, a very excellent grade for this purpose being the "Gold Label Styrian" brand. The reamers used in this operation are fluted, and at the works of the American Gun Barrel Mfg. Co. a double reamer has been used with remarkable success. This reamer removes as much in one cut as was formerly taken in two. The construction of the reamer is such that it is smaller near the shank than it is at the other end, the part toward the shank being so proportioned that it will re-

Fig. 9. Pratt & Whitney Reaming Machine.

move about 0.005 inch at one cut, while the second or finishing section removes 0.002 inch.

An illustration of a Pratt & Whitney gun barrel reaming machine is shown in Fig. 9. An enlarged view is shown in Fig. 10. The reamers are pulled through the work instead of pushed, as in drilling. The cutting edges of the first section of a double reamer are notched at regular intervals in order that the stock which is cut out of the barrel will not clog the cutting edges and cause the tube to "ring." "Ringing" a barrel means that under certain conditions a reamer drill or boring tool cuts a deep circular hole below the otherwise level inside surface, thus causing a serious defect. The first section of a double fluted reamer contains only one-half the cutting edges of the second half. Reamers are not hollow like drills, and instead of the oil passing through the centre of the reamer, it flows through the reamer shank into the barrel. The reamers do not revolve at the high rate of speed at which the barrels revolve while being drilled, but the operation of

Fig. 10. Enlarged View of Reaming Machine shown in Fig. 9.

reaming is performed much quicker than that of drilling, because the feed is greater. Six Pratt & Whitney reaming machines will take care of the output of ten drilling machines. Each reaming machine holds two barrels, and one man is required to run from two to three machines, while two men can run ten drilling machines.

The general construction of the Pratt & Whitney reaming machine will be understood by referring to Figs. 9 and 10. The drive is by a single pulley through gearing and a lead-screw, and a worm and bevel gears drive the two feed-rods. From these feed-rods the carriages are driven by bevel pinions and thrown out of gear by means of adjustable dogs. The head end of the machine is shown in Fig. 9 and reveals the bevel gears mentioned, together with some of the other details. In reaming, the barrels are clamped in the centre of a long oil pan shown in the illustration, and are held by the butts while the reamer shanks are run through the barrels by sliding the carriages which contain the oil pans towards the head of the machine. The reamers are then fastened on the ends of the shank and the oil-pan slide's run back until the cutting edges of the reamer come in contact with the barrels, after which the machine is started. The reaming speed for 0.303 calibre is about 200 r.p.m, and the feed is about 4 inches a minute. As a general thing, the roughing reamer has four flutes, while the finishing reamer has six. It is very important to keep the reamers in first-class condition if good work is expected, and the services of a skilled mechanic are therefore required for this purpose. Close watch is required, and the leeway allowed on finish-reaming is very small. An error of 0.00025 inch is sometimes sufficient to send the barrel to the scrap-heap, providing the hole is that amount too large. The final or

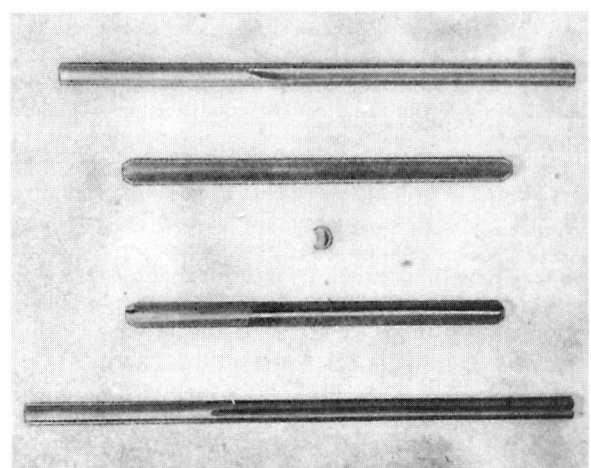

Fig. 11. Gun Barrel Drills and Reamers.

37

finish-reaming operation is done by practically the same method as that just described. Closer work and more skillful workmanship, however, are required on the finishing cuts than on the first cut, inasmuch as practically no leeway is allowed. On the finish cut all interior imperfections must be removed and the inside walls of the barrel must be made as clean and bright as the polished surface of a looking-glass, lest the keen-eyed inspector may discern some slight defect which would cause rejection of the barrel. Fig. 11 shows a group of drills and reamers used in the operations just described.

Reaming Gauges

The gauges used for the reaming operation are round and have a very slight taper. They are marked at intervals by fine lines which completely encircle them, and which indicate fractional parts of a thousandth of an inch. On account of the closeness of the work demanded, these gauges are changed from time to time, and in the larger and more important small-arms factories they are taken up weekly in order to discover and remedy any inaccuracies. The manner in which the reaming gauge is used is clearly indicated in Fig. 4, which shows an operator testing the accuracy of a barrel that has just been reamed.

Forming and Turning Barrels

After the reaming operation, the barrels are taken to a turning lathe where they are spotted in the centre preparatory to putting them on the barrel turning lathe. "Spotting" means that a cut is taken in the centre of the barrel about 1 inch in length and to the diameter to which the outside of the barrel is to be first turned. The centre-rests on the barrel turning lathe are brought up against this spotted centre while the turning processes are in progress. The operation of spotting is clearly shown in Fig.

Fig. 12. Spotting Centre of Barrel before Turning.

Fig. 13.
Turning and
Forming
Barrel

12. In turning the barrel after the spotting operation, the butt end is held by a dog while the muzzle is fitted to a centre plug and the rest at the centre of the barrel steadies it as indicated in Fig. 13. The two tool carriages, each of which carries a cutting tool, are now adjusted and each tool is started at the same time on the barrel, one at the muzzle and the other at the spotted centre portion. The tool carriages feed toward the butt end of the barrel and remove about one-eighth inch of stock during the turning operation. Over each cutting tool, a flexible hose is adjusted, through which a stream of soda water and oil is directed onto the cutting edge of the tool, lubricating and cooling it at the same time. Another stream of lubricant plays on the muzzle as it revolves on the centre. Automatic adjustable knock-offs are provided so that when the two tools have reached the limit of the cut the feet are knocked off.

Forming Barrels

All rifle barrels are not of the same shape, although military barrels of today are very similar. Thus in barrel turning the turning lathe is equipped with formers of various shapes which conform in outline to the shape of the barrel to be turned. Fig. 14 shows a group of these forming plates, which are adjusted at the back of the lathe in such a manner as to act as guides for the tool carriage. In the majority of cases, the lathes used in these forming operations are equipped with two carriages as mentioned, although in some factories four or more cutting tools are used in order to increase the production as far as possible. The operation of turning and forming the barrels is clearly shown in Fig. 13. As a general thing, about one-eighth inch is taken off from a barrel at one cut, and about fifteen minutes is the average time required for turning. It must be understood that nice

Fig. 14. Forming Plates used for
Barrel Shapes.

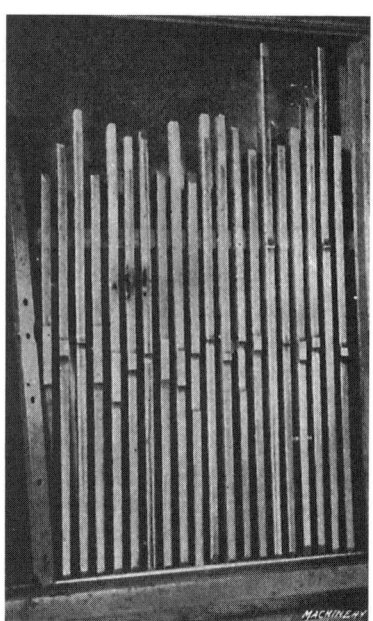

work is required in this operation, as the exterior surface of
the barrel must be put in the
best condition possible for the
grinding operation. Grooves of
any depth are not permissible,
and the turned surface must not
show roughness to any extent.
The cutting lubricant used for
the turning and forming operation is generally a composition
of soda water, although some of
the compounds on the market
are often used as a turning lubricant. The speed of turning
varies from 450 to 500 revolutions per minute and the rate of
feed is about one inch per
minute. It is of the greatest importance that the cutting tools
used in turning should be very
sharp, for there is nothing that
will crook a rifle barrel more easily than dull cutting tools.
Oftentimes the barrels are
sprung through the use of dull
cutting tools, so that when the
barrel straightener has made the
inside of the barrel straight and
true, it is found that the outside
is badly crooked and entirely out
of shape.

Straightening Barrels

The straightening of the barrels is conceded to require the
greatest skill of any operation in
the manufacture of rifle barrels
As the writer of this article has
spent nearly a third of a century
in the occupation of straightening barrels, and as there has
probably never been a compre-
hensive article written on this
subject before, the operation will
be treated exhaustively here in
the hope that a real understand-
ing of the method of rifle barrel
straightening may be conveyed
to the minds of the readers.

The first requirement for a
rifle barrel straightener is the
keenest and most perfect eyesight; in fact, there must be posi-
tively no defect of any kind in the
eye. One of the most peculiar fea-
tures in connection with this is
that no one knows whether or
not his eyesight is sufficiently
perfect for this line of work until
he has tried to straighten rifle
barrels. Young men who have
supposed their eyesight to be
perfect, and who have never suf-

fered from eye trouble of any kind, have endeavoured to learn the art of rifle barrel straightening and have found that within a short time their sight has greatly deteriorated, so that they have been obliged to withdraw from the work in order to preserve their eyesight.

It is not generally known that but one eye is trained in the art of barrel straightening, and in the writer's long experience as a straightener he has met but one person who could straighten barrels by using either eye. To the average straightener, the inside of a rifle barrel looks as strange to his untrained eye as it does to a man who is not a straightener, and in order to straighten barrels by using the untrained eye time would be required in which to train it to do the work. Despite the fact that only one eye is used in the straightening process, the barrel straightener never squints either eye, but views the interior of the barrel with his trained eye while both eyes are wide open. It is amusing to a barrel straightener to see a person squint one eye in order to look through a rifle barrel. In connection with the straightening of rifle barrels, it should be said that there is no one living who can tell by the glance of an eye through the interior of a rifle barrel whether or not the barrel is true so well as a straightener. The barrel straightener also learns by constant practice and training just how much can or cannot be done to a barrel.

In straightening a rifle barrel, the three important tools required are a shade, a straightening block, and a hammer. The best light possible must be provided in order to produce the best results. Straightening shades should be placed in windows having northern exposure which ensures a steady light all day long and eliminates those shadows that fall across the shade from surrounding objects when other exposures are used. The ideal arrangement for rifle barrel straightening is that in which the window shade consists of a ground glass window pane, one half the size of a window, and fitted into the upper half of the sash. Across this ground glass pane, about in the centre, a stick is fixed horizontally. This stick varies in thickness according to the distance of the straightening block from the shade. Thus, if the block is located a considerable distance from the shade, a heavier stick is used than if the block is near the shade. At a distance of about twenty feet, the stick would be one-half inch in thickness in order to produce the necessary shade result in the barrel, as it is the shadow or shadows cast by this stick on each side of the interior of the barrel which indicate to the straightener whether or not the surface of the barrel is true and straight. Sometimes two sticks are fixed across a shade, one heavier than the other. This is in order that a heavy or a fine shadow may be used, according to the preference

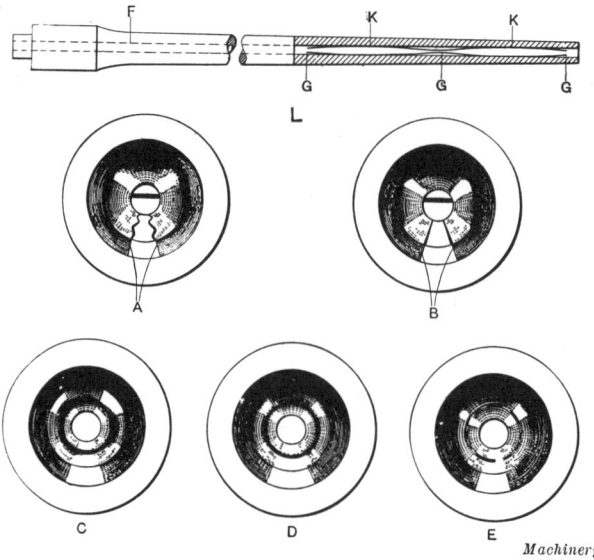

Fig. 15. Diagrams showing Appearance of Defects in Rifle Barrels.

of the straightener.

The shadows on the inside of the barrel are two thin, dark lines, one on each side of the interior surface of the barrel and slightly below the centre of the bore as indicated in Fig. 15 at *B*. These lines extend from the ends of the barrel nearest the shade to and slightly beyond the centre, after which the shadows diffuse somewhat. To one unfamiliar with the interior of a barrel, the lines appear to extend back from the ends of the barrel only an inch or two, and practice is necessary in order to familiarize the eye with their appearance.

Now, if these lines run straight, as shown at *B*, it indicates that the end of the barrel where they are located is straight. However, should they diverge from a straight line and zigzag in various directions as the barrel is revolved by hand, as shown at *A*, it indicates that there are crooks in the barrel. When the appearance of these lines is anything but straight, the straightener is obliged to use his skill in overcoming the defects, and when the shadows are finally made to run straight on one end of the barrel the other end can be treated in like manner. When the shadows run straight on both ends, the barrel may be considered as being straight from one end to the other in its entire length.

The straightening block shown in Fig. 16 consists of a heavy base on which are fastened two narrow steel dies which converge at the left side of the operator. These dies are about 1/2 inch in width and of hardened steel, in order to withstand the severe service to which they are put. The object of locating them in vee form is in order

42

Fig. 16. Straighten- ing Rifle Barrel.

that the spot in the barrel where a long crook is indicated by the shadows may be so placed that it will come about right on the dies. The reason for this is that the barrel must be straightened for a long distance when a long crook is encountered, and for only a short distance when the crook is short and sharp. The barrel is placed on these dies, while it is struck directly between them at a spot nearest the centre where the first crook is located. The hammers used in straightening average about 5-1/2 lbs. in weight and are of mallet shape. Those used for rough-straightening, prior to the turning operation, but after drilling, are of steel, while those used after turning are of copper or babbitt.

The copper hammers are to be preferred, because of their greater durability, and also on account of the greater sensitiveness which the copper hammer gives, permitting the straightener to gauge his blow to a nicety. In straightening a barrel, if it is struck at just the right place and with just the proper weight of blow, the crooks will be eliminated, and the straightener can go ahead to the next defect, working always away from the centre of the barrel towards the end.

It is absolutely necessary that the middle crook be eliminated before the next one can be operated upon successfully. In other words, each crook must be got out in its turn, beginning at the centre of the barrel and working towards the end farthest from the operator. There are several methods of shading by means of which barrels are straightened, but the method which this article describes is one which is recognized as being the most advanced. The shadows in a crooked barrel approach each other from each side of the interior of the barrel in places and diverge in other places. The point at which the shadows approach each other is

43

the proper place for the hammer blow to be struck, directly between the lines. If the blow falls just right, the line will spread and run true at that point, which will indicate that that particular crook has been eliminated. A little farther on in the barrel it will probably be found that the lines diverge or recede from each other, and it is important that the barrel should never be struck at this spot, for a blow between the diverging lines would cause them to diverge still more, which would mean that the crook would be intensified. By spinning the barrel round between the hands in the upright rest which is fixed adjacent to the operator, it will be found that directly opposite the place where the lines diverge they approach each other. Thus we have the same condition under which the first crook was eliminated, and a blow directed between the lines at this point, if properly struck, will also cause the lines to spread here, so that the second crook will be overcome.

This same method is followed to the end of the barrel, and when all these crooks have been removed and the shadows run straight from the centre to the end of the barrel, the other end may be operated upon.

Those who are not barrel straighteners are frequently fooled by the appearance of the shadows in the interior of rifle barrels when the surface is uneven. In this case they do not run straight, but in some places apparently run out of a straight

line. No matter how the barrel may be turned and moved, this condition remains the same. In other words, the shadows look the same all round the interior surface of the barrel. The skilled straightener knows just what this indicates, and he is also aware of the fact that all the pounding in the world cannot improve such a barrel in the least, for only reaming can eliminate this crooked appearance of the lines. In a case of this kind, if there should be enough stock in the barrel to permit a perfect levelling and smoothing up of the surface by means of reaming, it would be found that the lines would assume a straight and true appearance. This matter of uneven surface of the interior of a rifle barrel is one of the most unsatisfactory problems with which the barrel straightener has to contend, as the inspector, who is not a barrel straightener himself, decides that the barrel is crooked on account of its appearance, and therefore sends it to the straightener to be straightened, when in reality it should have been returned to the reamer.

The difficulties of the rifle barrel straightening vary with the calibre of the barrel and the quality of the steel used in its manufacture, the soft steel barrels being much easier to handle than those of the harder steel, such as 0.45 carbon or the regular nickel steel. The smaller the calibre of a barrel, the more difficult it is to straighten, and when small calibre and nickel steel are combined, then, indeed, the straightener has his hands full in getting out a satisfactory day's work. Perhaps there never was a more difficult barrel to straighten than the 0.23 calibre, nickel steel, Lee barrel which was made for the U.S. Navy at the time of the Spanish-American War. Added to the natural difficulties was the Government inspection, which was exceedingly rigid and severe. There are few military rifles, however, now used by any Government in which the calibre is smaller than 0.30.

Straightening Machines

Before proceeding with a description of the hand straightening operations, it may be advisable to state here that in some of the principal gun concerns, straightening machines are used instead of the hammer and block. The writer has tried this method, and can testify that excellent work can be done with the machine, but it is not nearly as fast as the hammer and block method. The impression prevails in some quarters that anyone can run one of these machines and straighten rifle barrels with it, but this opinion is erroneous in the extreme, inasmuch as no one but a rifle barrel straightener can use such machine, and while good work can be done in this way, just as good work can be done and is being done every day with the block and hammer, it is doubtful if quite as difficult straightening can be done on a machine as with the hammer

and block.

It is admitted by those who use both methods that machine straightening is not nearly as economical as hammer and block straightening, and that one man can do fully as much work by the latter method as three can do on a machine. Machine straightening for shot gun barrels has proved more satisfactory than it has for rifle barrels, inasmuch as the former are much thinner, and therefore can be bent more easily. Sharp, deep crooks in nickel steel barrels cannot be removed with any degree of swiftness by a machine and not nearly as satisfactorily as with a hammer and block. That fine work can be done on machines on barrels containing moderate crooks, however, is admitted by all experienced straighteners who have used them.

When barrels are straightened with a steel hammer after such operations as turning and grinding, hammer marks are invariably left on the interior, these being finally removed by reaming. When a barrel has been properly straightened on the inside, the outside should conform and be perfectly true and straight, unless the turning has been improperly done. In order to discover whether or not the barrel is straight on the outside after it has been finished on the inside, it is placed in the centring machine and spun rapidly with the hand. If true and straight, the barrel will run true, but if the barrel should wobble during this test it is evident that the outside is not true, and if very bad it is rejected and scrapped.

Defects in Rifle Barrels

Many times the interior surface of a barrel is marked by a ring caused either by the drill or reamer wearing a circular hole in the surface. Sometimes the defects are slight, while at other times they are very pronounced. When the rings are slight, a reamer cut will generally clean them out, but when too deep to be eliminated by reaming they are "set in" by the straightener in such a way that the final reaming will remove every trace of the defects. "Setting in" a ring means that the straightener locates the exact spot of the ring, and by hammering completely round the barrel at that spot while the barrel rests directly on the die he raises the ring above the level of the interior surface so that when the reamer cut is taken only the raised surface containing the ring is cut out and the defect is thus removed.

The straightener soon learns the difference between a reamer ring E, a drill ring D, and a powder swell C, shown in Fig. 15. The reamer ring is invariably a succession of nicks or gouges which completely encircle the interior surface of the barrel at the spot where they are located. On the other hand, the circular cut of a drill is likely to be clean, deep, and startlingly pronounced. A powder swell is indicated by the "downhill" like appearance of its nearer and far-

ther edges which dip gradually to its extreme depth directly in its centre. The edges of the swell are always smooth, and the general appearance of the swell itself is smooth. This powder swell is produced during the testing of a rifle barrel, and is really a failure of the barrel to stand the strain of the test. The "proof-test" is always made before the barrel is rifled, and the powder charge is much heavier than that used in regular service.

One of the greatest difficulties with which the straightener has to contend is the variation in some of the stock from which rifle barrels are made. In places the stock is found to be hard, while in other places soft spots occur. The result is that the straightener gauges his blow on a crook as he judges it should be, but in the event that there is a soft spot at that particular place in the barrel, the blow is too heavy, with the result that the barrel is made more crooked than ever. A condition of this kind always results in much hammering that could have been avoided had the operator known of the soft spot beforehand. On the other hand, when a hard spot is encountered, no damage is done, as the blow is not heavy enough to cause the barrel to assume a more crooked shape. Hence, the crook is not disturbed, and the next time a harder blow is struck which may be just hard enough to make the crook disappear. There is no way of discovering this irregularity of stock until the hammer falls upon the imperfect spot.

The upper diagram *L* in Fig. 15 represents a barrel with one-half perfectly true, as indicated by the dotted lines at *F*, and the other half with crooks at *K*, which are shown somewhat exaggerated in the sectional view. When straightening crooks, the barrel must always be struck at intermediate points such as *G*, but never at points *K*, as this would intensify the defect.

Final Straightening

The straightening thus far described is done on barrels between the various operations mentioned, and a steel flat face or nearly flat face hammer is used. It is the straightening between the operations mentioned which is most likely to crook the barrel, and which takes place before the barrels reach the chambering or rifling stage in the manufacture. It is generally considered that there are no operations hard enough to crook barrels after they have passed beyond the grinding stage, but it has been found that barrels are crooked at times even after they are rifled, and that because of thus being crooked they fail to shoot as accurately as they should.

Such barrels have, as a general thing, also passed the browning operation, and in handling they must not be bruised in the least, either externally or internally, so that it is an exceedingly difficult thing to straighten the barrels under these conditions; yet a skillful straightener can put them in first-class condition even when they have reached this stage.

Such barrels are straightened on copper dies, a rawhide or copper hammer being employed. The barrels are wrapped in paper which is glued in several layers round, but not to the barrel itself. The blow of the rawhide hammer on the paper leaves no mark on the barrel, either inside or outside, nor do the copper dies bruise the barrel, so that when carefully done it is found that the defect has been overcome and that the barrel is once more in good condition. The opinion is general among first-class straighteners that all rifle barrels should be looked over in their finished state, and that those which need retouching should be attended to in this way before they are sent away from the factory to the market.

Grinding Barrels

When the barrels have been properly straightened, after the finish-turning, they are send to the grindstone to be ground, and after the grinding operation they are reamed again, after which they are ready for the finish-straightening. If properly ground, the straightener has simply to touch them up and correct any slight imperfections that may be found. They are then sent to the finish-reaming machine which cuts the interior surface to the exact size required before the barrels are sent to the rifling machine. Two methods are now in use for grinding the

Fig. 17.
Grinding
Outside of
Rifle
Barrel.

barrels. One is that in which large stones of great weight are employed and the other is by means of automatic grinding machines. Each of these methods has its advantages, but there is no doubt that barrels can be ground more expeditiously on the large stones than on the smaller ones of the grinding machines, especially when the grinder using the large stone is an expert at his work. When this is not the case, however, the barrels may be badly crooked both externally and internally, which makes it necessary for the straightener to do considerable work in order to bring them back to condition once more, and at the same time makes much work for the filers in filing out unnecessary hammer marks and other defects on the outside surface of the barrel.

The grindstones generally used in the operation of grinding rifle barrels are from five to six feet in diameter and from twelve to fourteen inches face width and weigh from 3500 lbs. to 2 tons each. They are swung on an iron frame of great strength, and when in operation revolve at about 185 r.p.m. Fig. 17 shows an operator engaged in grinding a rifle barrel, using a steel rod passed through the bore of the barrel from the butt toward the muzzle. The butt end of the rod tapers which causes the rod to grip the barrel at this point. The rod is equipped with a handle on which a hollow wooden grip is fixed. This handle is grasped by the operator who places the barrel in a shoe set parallel with the horizontal axis of the grindstone and distant an inch or so from the face. The grinder now presses the long lever with his hip, which forces the shoe toward the face of the stone by means of a toggle action, and as the barrel touches the stone, the grinder turns the handle of the rod on which the barrel is held so that it is revolved against the face of the stone in the opposite direction to that in which the stone is revolving. The barrel is always ground from the centre toward the butt, then from the butt toward the muzzle, working it across the face of the stone until it has been ground for its entire length. During this operation, a steady stream of water plays on the surface of the stone, which prevents the barrel from burning, and also prevents the stone from glazing. Furthermore, the water makes the stone cut faster by acting as a lubricant.

Grinding the exterior surface of rifle barrels removes all the defects and imperfections left by the turning process as well as the hammer marks caused by the various straightening operations which have preceded the grinding. An expert rifle barrel grinder has acquired through long experience a sense of touch which tells him the amount of pressure necessary to use in bringing the barrel into contact with the stone. A careless grinder, however, may press the barrel too hard against the stone and thereby cause it to buckle and bend. The average amount of stock removed from the barrel during the grinding operation is about 0.010 inch. A snap gauge and gauging board are used to determine the proper size for a barrel when finish-ground, the barrel being laid on the gauge board and gauged at the butt, middle and muzzle. After the barrels have been ground to their proper size, they are washed in limewater to prevent rusting.

As the face of a grindstone wears down, hard pebbles are often encountered embedded in the surface of the stone unless these are removed, trouble is likely to be caused by these stones marking the outside surface of the barrel. As soon as a pebble is discovered, the stone is stopped and the pebble cut out by the grinder, using a cold chisel and hammer in the operation.

Military rifle barrels may be made either with or without forged front sight lugs. In some factories, muzzle sight lugs are brazed on the barrel just before the finish-reaming operations, between finish-reaming and finish-straightening. There is a difference of opinion as to which method is the better, as the brazing on of these lugs is likely to crook the barrel, thus making it more difficult to straighten in the final straightening operation. The writer is in favour of the latter method, however, inasmuch as the barrel is in better shape to straighten at the muzzle through all the first operations when the muzzle is free of a sight lug than when the lug is forged on. The hardest part of the straightening is always found in the early operations on rifle barrels, and many times crooks are met with directly under the forged lug which are almost impossible to eradicate because the lug is in the way. When, however, the brazing of the lugs takes place just before finish-straightening, the crooks are much easier to knock out than if they have been encountered during the earlier operations.

Inspection before Rifling

After the rifle barrels have been finish-reamed, they are subjected to the final inspection for reaming and straightening, consisting of the following operations: interior gauging, leading, shading and surface inspection. The gauging operation is for the purpose of determining whether or not the bore of the barrel is true to the size required, as the slightest variation one way or the other may mean the rejection of a barrel, especially if it is over size. If under size, however, it is sent back to the finish reamer who reams it to the exact size. The leading operation is done by pushing a lead bullet through the barrel by hand, a brass rod being used for this purpose. In the slow passage through the barrel, the bullet which fits the bore snugly must not meet with any obstruction from one end to the other. Should there be any resistance, even the slightest, the barrel is rejected and returned to the finish reamer, who removes the imperfection. Just so is the barrel rejected in the shading operation, which determines whether or not the barrel is crooked or straight. When found to be crooked, it is up to the straightener to retouch and make the barrel straight. The slightest defect on the interior surface of the rifle barrel, such as rings, poor surface, gouges, scratches or markings of any kind whatsoever, which can be discerned with the naked eye, will cause its rejection and it will be returned to the finish reamer or straightener for corrections.

Rifling Barrels

After the barrels have passed the various inspections for workmanship successfully, they are passed along to the rifling machine where the grooves are cut which cause the bullet to rotate in its passage through the barrel. The number of grooves in rifle barrels varies from four to six or even seven in some instances. The twist also varies in different makes and models of barrels, some making a complete turn to every ten inches, and others requiring thirty six inches in order to make a complete turn. The operation of rifling is done on specially constructed machines and will not be described in detail in this article.

Other Operations on Barrels

After the barrels have been rifled, they are again inspected and given a thorough leading in order to remove all rough edges and smooth down the grooves. In some factories the operation is done by means of tools worked by hand, while in other factories machines are used. Whether done by machine or hand, the operation is the same, as a lead slug which is cast to fit the bore of the barrel is worked backward and forward through it, after it has been charged with fine emery and oil. After the leading operation, the chambering, threading, slot-cutting and extractor-cut operations are done. In the chambering operation, five tools are used; a counterbore, two roughing reamers, a heading tool and a finish-reamer. A "Ball" speeder is then used to take out the burr at the beginning of the rifling grooves, bevelling them to the right shape just where the bullet protrudes from the end of the barrel. In the finish-reaming operation on the chamber, the work is done by hand in some factories and on a chucking machine in others. The workmanship required in the cutting of the extractor slots must be exceptionally good, because a fractional part of a thousandth of an inch out of the way may mean spoiling a barrel. Poor workmanship here would mean that the cartridge might not fit as it should in the chamber or that after the rifle had been discharged the shell could not be extracted.

The slot-cutting operation is sometimes done before rifling, and the threading operation on the butt of the barrel is almost invariably done after rifling. The extractor cuts are made after chambering, and the sight studs are machined before rifling. Then comes the muzzle finishing, which is done partly by machine and partly by hand, the barrel being left longer in the rough state than is really necessary, in order that there may be no mistake in the final finishing.

Polishing Barrels

Military rifle barrels are polished on wheels which run at very high speeds, the larger wheels running at the rate of 3500 r.p.m., while the smaller ones attain an even higher speed than this. The wheels used for polishing have a wooden centre and are covered with leather, the leather being coated with glue on which a heavy coating of emery is sprinkled. For rough polishing, No. 60 and 70 emery is used, and for finishing No. 90 is used. Plugs or handles made of metal are stuck into the muzzle and butt of the barrel when it is ready to be polished and fit snugly into the rifle grooves so that the barrel will not turn on the handles and injure the rifling. The handles spin in the hands of the operator at the same high rate of speed at which the barrel is spinning, but as these handles are very smooth the operator's hands are not injured by the friction. At one time the work of polishing was done exclusively by men, but at the present time women as well as men are employed in this part of the work. After the barrels are polished, they are ready for the browning operation, which is the final stage through which the barrel passes before it is assembled.

All military rifle barrels are subjected to a high powder test before, and a target test after, rifling. The first test varies according to the grade of steel in the barrel, high nickel steel barrels being tested with a powder charge which would burst a low-grade steel barrel. The test charge includes a heavy leaden slug, and the combination is from two to three times as heavy as the charge which would ordinarily be used in service. Should there be seams in the stock or other defects, the first test preceding the test after rifling, which is made with the regulation charge in order to determine the accuracy of the rifle, will bring them to light.

Browning Barrels

The final operation to which rifle barrels are subjected before they are ready for the assembling room is that of browning. While the finished barrel is blue in colour and not brown, the operation is called "browning" because the barrels have to be rusted on the exterior surface before they can be blued. The coating of rust is obtained by covering the outside of the barrel with a browning mixture which causes it to rust after it has been baked in an oven heated by coil steam pipes. The air in the oven is moistened by perforated steam pipes through which the steam is allowed to escape into the oven. Before the rifle barrels are run into the hot oven for baking, they are subjected to a number of other operations, the first of which is wiping. In this operation the barrels are thoroughly wiped in order to relieve them of every particle of grease or other foreign substances and are then placed in circular iron trucks which are made to hold fifty barrels at a time. The truck and

barrels are then placed in an upright caustic soda tank where they are boiled for about fifteen minutes. From this tank they are taken to another, where they are thoroughly rinsed so that every particle of grease and oil is removed and the barrels are ready for the spongers who apply the browning solution.

It is well to state here that the most successful rifle barrel browners have their own secret formulae which are jealously guarded so that only the browners themselves know the exact proportions of the mixtures. It is no unusual thing, therefore, to find barrel browners who have held their positions for many years, on the strength of their knowledge in this respect. One of the oldest and most satisfactory formulae for browning rifle barrels consists of the following mixture:

Spirits of wine .... 5 ounces
Spirits of nitre .... 8 ounces
Tincture of steel .... 8 ounces
Corrosive sublimate
.... 4 ounces
Blue vitriol .... 4 ounces
Water .... 1 gallon

The barrels are kept in the oven for four or five hours and when taken out are heavily coated with rust. They are then immersed in a third tank of chemicals for fifteen minutes, a secret formula often being used at this stage. A preparation which will give excellent results, however, is as follows:

Tincture of muriate of iron
.... 1 ounce
Nitric ether .... 1 ounce
Sulphate of copper ....
.... 4 scruples
Rain water .... 1 pint

If the process is to be hurried, two or three grains of oxymuriate of mercury can be added. When the barrel is finished, it is placed in limewater for a short time to neutralise any acid which may have penetrated. The barrels are now ready for carding, which is done on a 14-inch wheel having a face 3 inches wide covered with carding cloth. This wheel revolves at about 1600 r.p.m. and the man who does the carding stands at the back of the wheel and presses the barrel against the rim. After the barrels are carded they are returned to the spongers for a second coating of the browning solution. The operations of boiling and scratching on the carding wheel are continued until all the pores of the steel are thoroughly coated and their condition is satisfactory to the inspectors, who give them a most rigid examination. After using the scratch brush, the following formula can be used:

Shellac .... 1 ounce
Dragon's blood .... 25 ounces
Rectified spirit .... 1 quart

Or:

Nitric acid (specific gravity, 1.2)
.... 1 part
Nitric ether .... 1 part
Alcohol .... 1 part
Muriate of iron .... 1 part

The above ingredients are mixed together, after which 2 parts of sulphate of copper and 10 parts of water are added.

After the barrels have been passed by the inspector, they are taken to the assembling room where they are assembled with the other parts. Many persons suppose that the browning of a barrel is done for the sake of appearance, and while this is partly the reason, the principal object is to preserve the barrel and prevent rusting. Were it not for the care taken in the browning operation, rifle barrels would not last nearly as long as they do.

Machining Rifling Bars
On The Bench Lathe
By A.H.C.

MACHINERY MAGAZINE - JUNE 8, 1916

It is the purpose of this article to describe the work of machining rifling bars on the bench lathe. The bars were made of steel, and the work involved deep-hole drilling and eccentric turning and grinding operations. It is not claimed that these bars could not have been machined in some better way, but it happened that the only equipment available at the time was three bench lathes, and the results obtained with these machines were very satisfactory. It is hoped that the following description may prove of value to readers of Machinery in suggesting ideas for handling bench lathe work where similar operations have to be performed. In the accompanying illustration the rifling bar is shown at *A*; the cutter is inserted in this bar at *B* and provision is made for adjusting the radial position of the cutter by means of a rod which has a tapered side that engages the bottom of the cutter. This rod is carried in the hole *C* in the rifling bar, which is 4-7/8 inches deep by 1/8 inch in diameter. The finished size of the bar is 0.300 inch in diameter, and the hole *C* is 1/16 off centre. It is the purpose to describe a fixture designed for drilling this eccentric hole, the work of drilling the hole, and the way in which the final turning and grinding operations were

performed to bring the outside of the bar parallel with the hole and still maintain exactly the required eccentricity.

As previously stated, three bench lathes were available for use in machining the rifling bars *A*. Before starting work, these machines were carefully tested and it was found that the beds were comparatively straight, but that the tailstocks were approximately 1/64 inch out of alignment with the headstocks. The following expedient was adopted to correct this inaccuracy. Chucks *D* were made with tapered shanks to fit the tailstocks, and these chucks were provided with thumb-screws *E* to hold the tools. The chucks were next mounted in the tailstocks of the respective bench lathes on which they were to be used and marked in one position, and the holes were drilled and reamed to receive the sockets *F* by means of tools held in the headstock chuck. Brass sockets *F* were next made for each of the drills and reamers that were to be used in drilling the holes *C* in the rifling bars.

The fixture for holding the bar while drilling the hole *C* is shown at *G* in the accompanying illustration; and in passing it may be mentioned that the same type of fixture may be used to advantage on the various

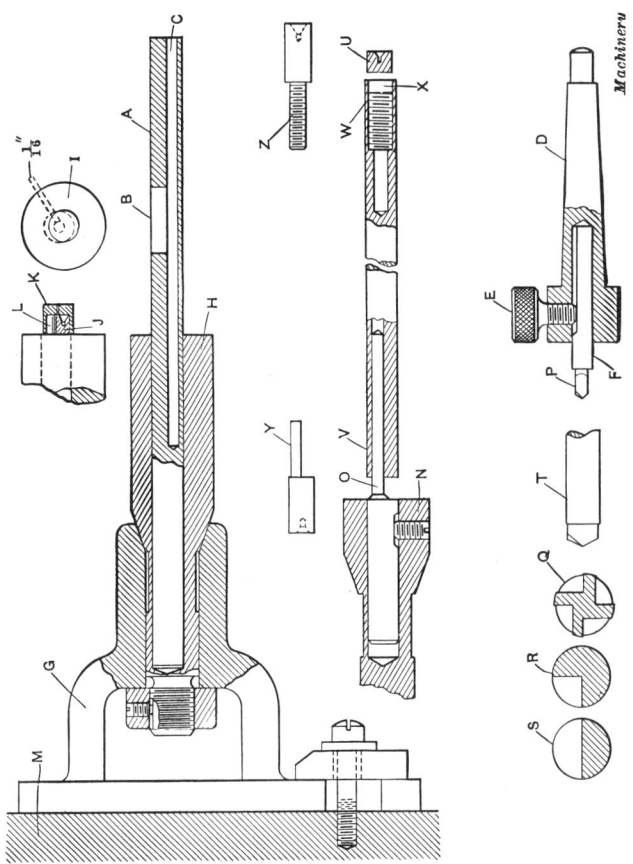

Tools used for drilling, turning and grinding Rifling Bar on Bench Lathe. Tools Q, R, S, and T are shown to an Enlarged Scale.

classes of lathe, drill press and milling machine work. In the present case the fixture was also used for holding the chuck *H* in the proper position to drill the eccentric hole to receive the rifling bar *A*. The method of locating the chuck in the fixture preparatory to drilling the hole was as follows. The required eccentricity is 1/16 inch, as indicated in the end view *I* of the chuck. To provide for obtaining this eccentricity, a projection *J*, 3/16

inch in diameter, was turned on the end of the chuck body, while the latter was held on centres so that it was concentric with the lathe spindle. A cap or shell *K* was next made of such a size that the difference between the inside of the cap and the diameter of the projection *J* on the chuck was 1/8 inch; i.e., twice the required eccentricity. After this cap had been made, a plug *L* was turned up to a diameter of exactly 1/8 inch and the cap. *K* was then forced over the projection J and plug L, as shown in the illustration. The fixture G was next mounted on the faceplate M of the lathe and its position adjusted until the centre of the cap K coincided with the centre of the lathe spindle, as proved by indication on the centre of the cap. The chuck H was then drilled and reamed to receive the rifling bars A which were held in place in the chuck by means of two set screws that are not shown in the illustration. After the chuck H had been completed, the next step was to make a second chuck N without altering the setting of the fixture on the faceplate. After this chuck had been finished, an eccentric mandrel O was turned up and hardened, after which it was remounted in the chuck N and ground to ensure accuracy.

The hole C was started with a short flat drill P which was followed by a twist drill that worked to a depth of 1 inch, after which a boring tool was used for machining the hole to a depth of 3/4 inch. A four-lipped reamer Q was next employed to finish the hole to the depth reached by the first twist drill. The accuracy obtained with these preliminary operations was highly important, because any error introduced up to this point would influence the accuracy of the entire job. In connection with the machining of deep holes of small diameter, it may be mentioned that the writer has never obtained satisfactory results with the so-called "cannon" drills of the form shown at

R or half reamers of the form shown at S. The corners on tools of these types are too easily dulled and too much resistance is offered to the clearance of the chips from the holes. But the reamer Q gave very satisfactory results; the ends of the lips of this tool are not rounded and they are backed off on the top just enough to give a free cutting action. A tool of this type is easily "stoned" on the ends of the lips to keep it in good working condition. The writer is prepared to recommend this type of tool for use in reaming deep holes of small diameters. After the hole had been finished to a depth of 1 inch, special No. 31 twist drills 1-3/4, 2-1/2 and 6 inches in length were used to complete the drilling of the holes to the required depth. These tools were used in conjunction with reamers of full length. If an exceptionally smooth finish had been required, a second reamer would have been used in connection with each drill and the lips of this second reamer would have been rounded.

The tool T is one of the most important of those used for drilling the holes. It will be seen that the body of this tool is slightly. larger than the cutting point; this body fits closely into the reamed hole and is of sufficient length to realign the other drills for a depth of 3 or 4 inches. Each of the No. 31 twist drills works to a depth of about 1 inch, and between successive operations of the twist drills the drill T was used to maintain the alignment of the hole by recentring. From three-quarters to one hour was required for drilling hole C in each of the rifling bars, and the writer drilled and reamed over 90 of these holes without breaking or choking a drill. The accuracy was sufficient to allow a piece of drill rod 0.0015 inch under size to drop to the bottom of the hole through the force of gravity. Two reamers were broken in reaming the ninety holes referred to, and the cause of this was that the tools were inadvertently started to work under conditions of speed and pressure suitable for the drills, which were too severe for the reamers.

After completing the drilling of the holes in all of the rifling bars, the next step was to finish the outside of the bars to the required diameter; and in performing this operation precautions had to be observed to ensure having exactly the required eccentricity for the hole and also to have the hole parallel with the outside of the bar. For this purpose the chuck N was mounted in the fixture – with the position of the fixture on the lathe faceplate unchanged – and the mandrel O was mounted in the chuck so as to ensure replacing in the same position, but in use, of course, it was placed in the lathe itself. The mandrel is of such a size that it is a wringing fit in the hole C, and in preparing to machine the outside of the rifling bars the first step was to mount one of the bars on the mandrel as shown in the illustration. It will be noted that the position of

the rifling bar on the eccentric mandrel is reversed in the lathe, with the result that the axis of the bar is brought concentric with the axis of the lathe spindle. When the position of the work had been adjusted so that the outside of the bar ran approximately true, the bar was soldered to the mandrel, after which a centring button U was brought near the opposite end of the rifling bar A by means of the tail-centre. This centre was then "sweated" on to the end of the rifling bar. In securing the centre in place, great care had to be taken to avoid deflecting the rifling bar, and after it had been soldered in place the work had to be tested to see that such deflection had not occurred. This was easily determined by backing the tailstock away from the work and rotating the lathe spindle with the outer end of the work free. Under these conditions any disturbance in the alignment of the work can be readily detected.

The next step was to turn the outside of the rifling bar at V and W to a diameter of 5/16 inch to form bearings for steadyrests. After this had been done, the work was released from the centre U and mandrel O by melting the solder; it was then mounted in a spring chuck and supported by a centre at the opposite end, after which the entire outside of the bar was turned to a diameter of 5/16 inch. The next step was to drill and tap the hole X, and for this purpose special taps were employed. These were made with an enlarged section between the tap and the straight shank which entered the opening in the collet *D*. The tap was turned by a pin entering holes in the large central part of the tap, the work being held in a chuck and further secured by means of a dog. After finishing the machining of the tapped hole *X*, the plugs *Y* and *Z* were introduced into opposite ends of the rifling bar and carefully centred while the bar was supported in steadyrests on the bearings machined at *V* and *W*. It will, of course, be evident that plug *Y* was soldered to secure it in the desired position. The work was then set up on centres on the grinding machine and ground while soft to a diameter of 0.306 inch, after which it was hardened, straightened and finish-ground to the required diameter of 0.300 inch.

Drilling A Long Blind Hole

MACHINERY MAGAZINE - JULY 4, 1918

W. H. J. – We have some long blind holes 6 m/m. diameter by 132 m/m. deep to drill in phosphor-bronze stick, as shown in Fig. 1. Would you do us the favour of advising us the quickest methods? We propose to machine this job by the bar on a capstan lathe.

A. – The required degree of accuracy of the drilled hole not being stated above, several methods by which the work may be accomplished will be discussed in this article, from which the querist will be able to choose the one most suitable to his requirements. The work is shown in Fig. 1, and in any case it will be necessary to divide the whole work into two separate operations. In the first operation the work will be turned and parted off from the bar in the usual way, but will not be drilled. The drilling will be done in another handling, and although the capstan lathe is not exactly the most economical machine tool to use for drilling deep holes, very little trouble will be experienced if reasonable care is taken to ensure the drills being accurately in line with the centre of the spindle.

The hole is 132 m/m. deep by 6 m/m. diameter, or approxi-

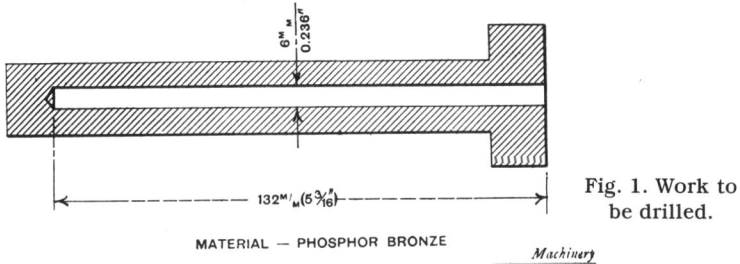

132ᴹ/ₘ(5 ³⁄₁₆")

MATERIAL — PHOSPHOR BRONZE

Fig. 1. Work to be drilled.

Machinery

mately 5-3/16 inches by 15/64 inch diameter. The writer has seen thousands of pieces having a hole of the same diameter and 3 inches long drilled in the turret lathe, using an ordinary centring tool, followed by twist drills, the material being, as in this case, phosphor-bronze. In the above case the drill was withdrawn from the hole after every half inch drilled, in order to clear the-drill of chips, and to carry lubricant inside the hole. The withdrawal of the drill is done very rapidly. The holes produced as above may not have been perfectly concentric, but the error was very small, and the scrap practically nil. Maybe the above method could be applied to the piece under consideration, and in any case it will be worth while to experiment a little, in order to prove whether or not the requirements of the case can be met in this manner. The chief consideration in deep hole drilling lies in having a true start for the hole; methods by which this essential feature is obtained will be considered next.

Greater accuracy in the concentricity and the size of hole will be obtained by the method indicated in Fig. 2. In this instance the work is centre drilled in the first operation, and in the second is drilled to a depth of about 1 inch, by means of a twist drill. In the third operation the drilled hole is bored by means of the single-pointed boring tool shown, and reamed in the fourth operation.

The sizes of the hole in these stages are shown in the drawing. The operations 1 to 4 are for the purpose of starting the hole perfectly concentric with the centre of the lathe spindle. In the fifth and sixth operations the hole is drilled to the required depth by the D drills, as shown. These drills made with a chip gap of one-quarter to one-half of the whole circle, and the body of the drill in each case is a sliding fit in the hole reamed in the fourth operation. The diameter is too small to allow for a channel all the way up the drill to provide a passage for chips and lubricant, and under these circumstances the drill will have to be withdrawn from time to time in order to clear the hole of chips. A copious supply of lubricant should be played on to the drill, and also forced up the hole when the drill is withdrawn.

A more elaborate method of drilling the hole, ensuring greater accuracy, is shown in Fig. 3. In this instance the first five tools are held in a Marvel chuck mounted in the turret. The tools are inserted, operated, and removed in the order shown, and are provided with stop collars causing them to project the correct amount from the chuck. Operations 1 to 4 are the same as in the previous method, but operation 5 consists of re-centring the bottom of the hole. This is followed by a twist drill, in operation 6. The hole is reamed in operation 7, and the bottom is again centre drilled in operation 8. In operation 9 another twist drill finishes the hole to the required depth, and the hole is reamed in operation 10. The employment of twist drills which are either of special length, or are standard drills soldered to shanks, is preferable from the drilling point of view, but where accuracy is essential the use of the reamer and centre drills is absolutely necessary in order to produce a hole which is straight and true to size. The utmost care should be exercised to ensure the drills and other tools being exactly in. line with the centre of the machine.

There is another way of drilling the hole, and one worth careful consideration. In this method the piece would be made as be-

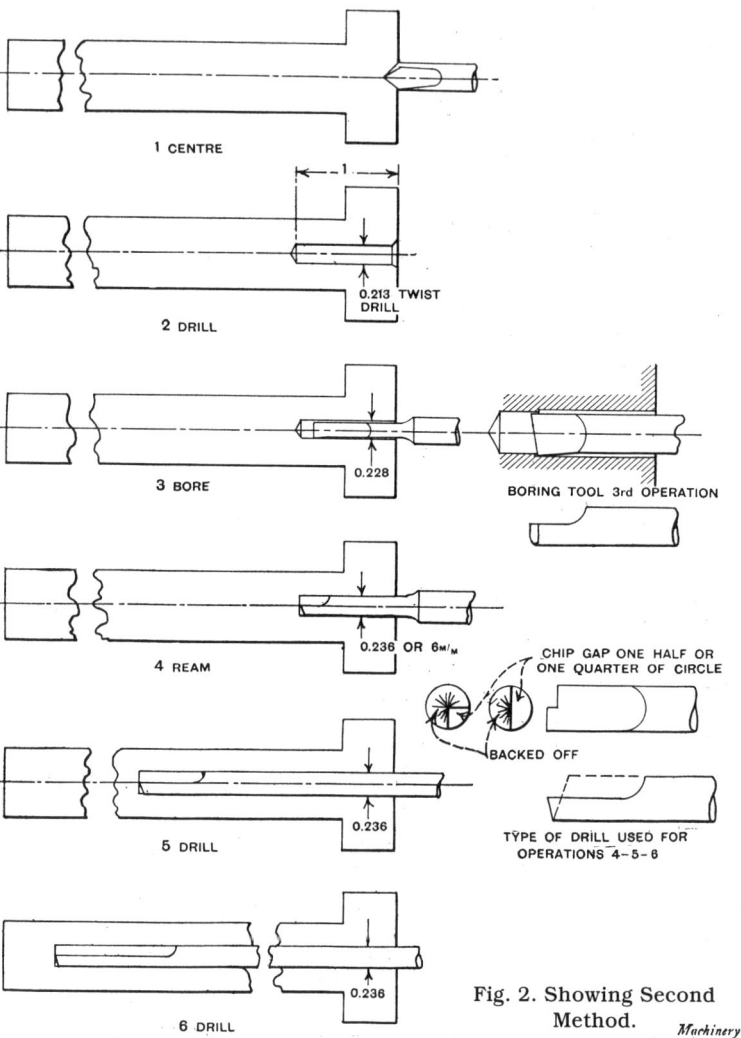

Fig. 2. Showing Second Method. *Machinery*

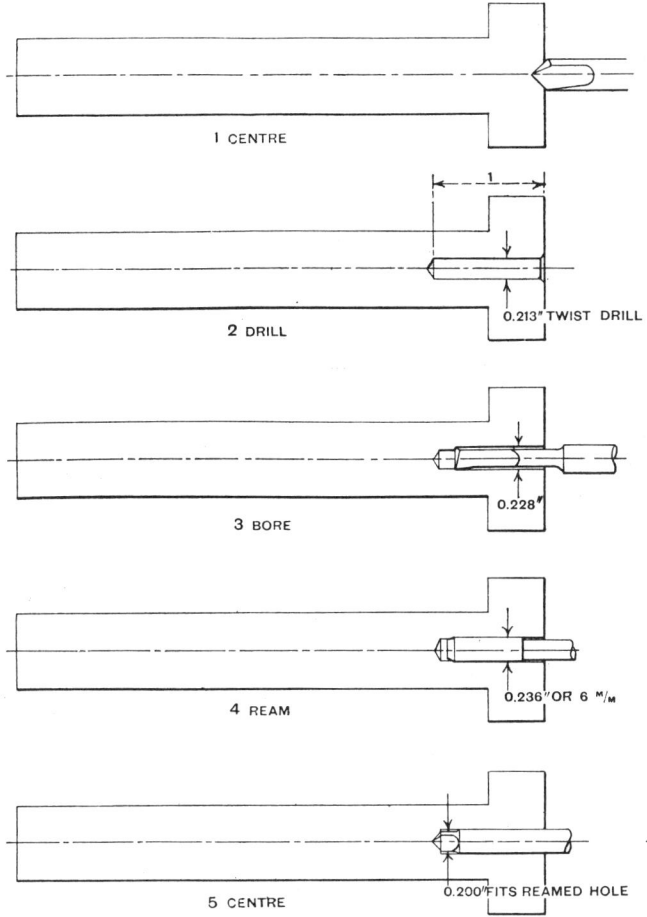

1 CENTRE

1

0.213" TWIST DRILL

2 DRILL

3 BORE

0.228"

0.236" OR 6 M/M

4 REAM

0.200" FITS REAMED HOLE

5 CENTRE

FOR OPERATIONS 1 TO 5 HOLD TOOLS IN "MARVEL" DRILL CHUCK MOUNTED IN TURRET

Fig. 3A. Showing Third Method.

fore on the turret lathe, but drilled on the drilling machine, holding the drill stationary on the machine table, and the work in a chuck attached to the machine spindle. The method, is indicated in Fig, 4, and, as will be seen from the drawing, a special long twist drills held in a Marvel chuck, which is fixed to a cast-iron base. The base, or slide is an easy fit in the bed of the fixture to allow it to float slightly; the slide may be moved towards or away from the operator to facilitate removal of the work from the chuck. If desired, the slide can carry a reamer side by side with the drill, and held in the same manner in another

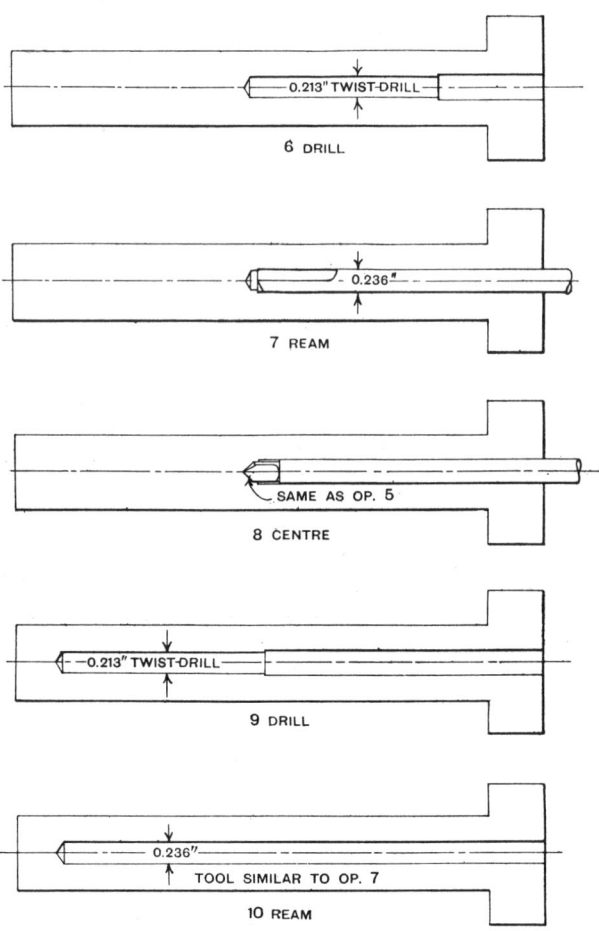

6 DRILL — 0.213" TWIST-DRILL

7 REAM — 0.236"

8 CENTRE — SAME AS OP. 5

9 DRILL — 0.213" TWIST-DRILL

10 REAM — 0.236" — TOOL SIMILAR TO OP. 7

Fig. 3B. Showing Third Method.

chuck, using this method, the work would be turned and parted off from the bar. An operation of centre drilling would follow, and at the same time the front face of the work could be faced. The drilling of the hole on the drilling machine would follow next. A small, three-jaw scroll chuck may be used on the spindle to hold the work.

Four different methods of drilling the hole have now been described, the first method of drilling by means of twist drills on the turret lathe, and the fourth by the same means on the drilling machine, would produce what may be termed in this instance second-grade accuracy. Of the two methods the fourth has the advantage in the way of

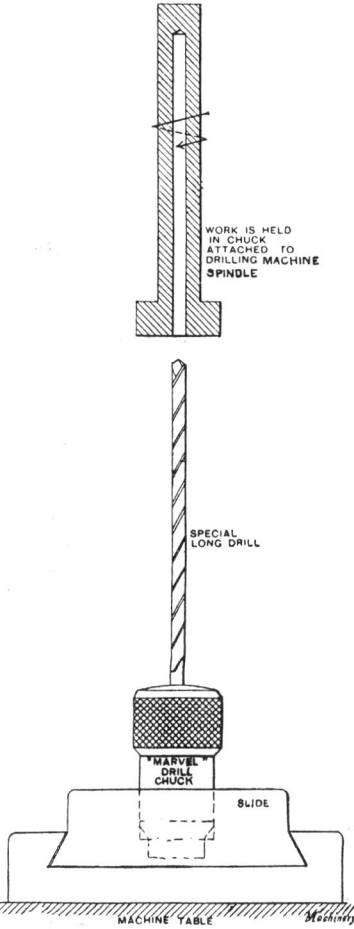

Fig. 4.
Showing
Fourth
Method.

WORK IS HELD
IN CHUCK
ATTACHED TO
DRILLING MACHINE
SPINDLE

SPECIAL
LONG DRILL

"MARVEL"
DRILL
CHUCK

SLIDE

MACHINE TABLE   *Machinery*

clearing-away the chips more easily, which fall by their own weight; a pipe carrying lubricant can be arranged to force a supply up the flutes of the drill, a pump being used for the purpose. A splash guard can be arranged to prevent the escape of the lubricant from the table of the machine.

The second and third methods will produce better holes. but will take a longer time than ei-ther the first or fourth methods. The second method, Fig. 2, calls for fewer tools than the third, Fig. 3, but in operation the latter should produce more accurate work. The method to be used depends entirely on the degree of accuracy required, and from the four schemes submitted herewith the querist will doubtless be able to select the one suitable to the requirements of his case.
W. R.